POLITICAL CULTURE IN EAST CENTRAL EUROPE

This book was edited under the auspices of the Austrian Institute for Eastern Studies, Budapest, Publication was sponsored by the Austrian Federal Ministry of Science, Research and the Arts.

Political Culture in East Central Europe

Edited by
FRITZ PLASSER
ANDREAS PRIBERSKY

Avebury

Aldershot • Brookfield USA • Hong Kong • Singapore • Sydney

Published by
Avebury
Ashgate Publishing Limited
Gower House
Croft Road
Aldershot
Hants GU11 3HR
England

Ashgate Publishing Company
Old Post Road
Brookfield
Vermont 05036
USA

British Library Cataloguing in Publication Data

Political culture in East Central Europe
    1. Political culture - Europe, Central 2. Europe, Central -
    Politics and government - 1989-
    I. Plasser, Fritz II. Pribersky, Andreas
    306 . 2 ' 0943

ISBN 1 85972 259 8

Library of Congress Catalog Card Number: 96-84013

Printed and bound in Great Britain by Ipswich Book Co. Ltd., Ipswich, Suffolk

Contents

Figures and Tables

List of Contributors

Attila Ágh
Attila Ágh is Professor of Political Science at the University of Economics, Budapest, Hungary.

Adolf Bibič
Adolf Bibič is Professor of Political Science in the Faculty of Social Sciences, University of Ljubljana, Slovenia.

Danica Fink-Hafner
Danica Fink-Hafner teaches Political Science in the Faculty of Social Sciences, University of Ljubljana, Slovenia.

Pavol Frič
Pavol Frič is Research Assistent at GfK-Institute, Prague.

Hans-Georg Heinrich
Hans-Georg Heinrich is Professor of Political Science at the University of Vienna.

Josip Kregar
Josip Kregar is Professor in the Pravni Fakultet Svevcilista Zagreb and Head of the Croatian Political Science Association.

Karin Liebhart
Karin Liebhart is Research Assistant at the Association for Political Information, Vienna, Austria.

Igor Lukšič
Igor Lukšič is Assistant Professor of Political Science, Faculty of Social Sciences, University of Ljubljana, Slovenia.

Gregor Matjan
Gregor Matjan is currently a freelance Social Scientist working in Vienna.

Gregorij Mesežnikov
Gregorij Mesežnikov is a member of the Slovak Academy of Sciences and teaches Political Science at Comenius University, Bratislava, Slovakia.

Silvia Mihalikova
Silvia Mihalikova teaches Political Science at Comenius University, Bratislava, Slovakia.

Pavle Novosel
Pavle Novosel teaches Journalism at the University of Zagreb, Croatia.

Emanel Pecka
Emanel Pecka is Professor at the University of Economics, Prague.

Fritz Plasser
Fritz Plasser is Professor of Political Science at the University of Innsbruck, Austria.

Andreas Pribersky
Andreas Pribersky is responsible for Social Sciences at the Austrian Institute for Eastern Studies in Vienna, and teaches Political Science at the University of Vienna, Austria.

Máté Szabó
Máté Szabó is Professor of Political Science at the University of Budapest and General Secretary of the Hungarian Political Science Association.

Peter Ulram
Peter Ulram is Assitant Professor of Political Science at the University of Vienna and Head of the Social Research Department (GfK-Institute, Vienna).

Preface

Fritz Plasser and Andreas Pribersky

Political jokes once passed for the essence of waywardness under East Central European Real Socialism, and are only gradually reappearing after they briefly ceased to serve as expressions of tongue-in-cheek obedience. When, for instance, a foreigner voices concern about the victories of "reformed communists" at the polls, and at nationalists coming out as the strongest opposition parties, locals try to allay his fears: "don't be afraid. Voters know what they're doing. They just don't know why." Despite grammatical appearances to the contrary, the quip ends with a question. The question, that is, of whether, how, and why East Central European political attitudes change: the question, of course, which has inspired this book.

Political culture and its development must be seen as one of the most important aspects of East Central European democratization. Along with changes in economic, legal or institutional frameworks and political systems in general, the historically formed and historically variable attitudes, evaluations and practices of politics have come to be seen as major indicators of transformation. Especially with respect to the countries featured in this volume – Croatia, the Czech Republic, Hungary, Slovakia, and Slovenia – renewed or incipient interest has of course also been motivated by these countries' geopolitical position at the European Union's "frontier", as well as in recognition of the fact that all of them are – though to varying degrees and with varying determination – aiming at a rapprochement and eventual full-scale integration into the Union.

In comparing and evaluating political culture, however, we encounter difficulties easily eschewed in any ranking of, say, inflation rates or economic growth data. Research on political attitudes and practices uses surveys on participation, political efficacy, or the credibility and support of political systems. But it must also address the complex relationships between such attitudes and practices, as well as a variety of societal developments instrumental in shaping them. The latter need in turn to be situated in the context of nationally specific factors of *longue durée*. Finally, "hard" data interpretation as well as "soft" historical tracings are themselves bound to the context to which they refer. But no matter how difficult it may seem to enter this thicket of complexities, the surveying and description of political cultures will undoubtedly enable us to give more precise and circumspect answers to questions concerning perspectives and likely scenarios of democratization in East Central Europe.

In compiling this volume, we have tried to take account of the diversity of such questions, as well as of possible solutions to problems of methodology. The introductory essays present comparative data on the region together with in-depth discussions of methods applied, perspectives for further research and possible alternative approaches.

In subsequent chapters, authors familiar with the field and working in the countries under scrutiny discuss their nations' political culture in one predominantly empirical, and one more descriptive essay. The latter, especially, aim at outlining the current national debates on the concept and content of political culture, and at an interpretation of historical patterns "from within". In an attempt to provide a contrastive definition of that geographically vague and historically contested notion of (East) Central Europe, the two final contributions in this volume discuss the political culture of Austria – as the Western country with perhaps the closest historical ties to the region – and Russia – as heir to the Soviet Union dominating that part of Europe up to the revolutions, velvet or otherwise, of 1989.

This volume in general and its country-centred approach in particular would not have been possible without the cooperation of various Central European Political Science Associations, resulting in a conference on the subject of this book in Autumn 1994. The meeting was hosted by the Austrian Political Science Association and admirably organized by Doris Vogl. We would also like to thank Adolf Bibič, Othmar Höll and Máté Szábo for their support in bringing about the conference, which will be followed by a meeting in Budapest in 1995, and one in Slovenia in 1996. Finally, our thanks are due to Herwig Engelmann for copy-editing, and to Gertrud Hafner for layouting the volume. Without their painstaking and often difficult labours, this book could never have appeared in print.

Fritz Plasser
Andreas Pribersky

Innsbruck and Vienna, December 1995

INTRODUCTION

1 Measuring Political Culture in East Central Europe. Political Trust and System Support

Fritz Plasser and Peter Ulram

Some theoretical and methodological considerations

Six years after regime transition, Eastern Europe's new democracies are finding themselves locked in a paradox: while the transformation to a market economy appears increasingly successful and reform countries are proving more and more competitive economically, there is far greater ambivalence concerning the first available evaluation of political and institutional transformation processes (Nelson 1995). Naive hopes for a continuous and linear unfolding of democracy in Eastern Europe have been disappointed by unexpected backlashes, new instabilities and increasing tensions within, as well as between, reform countries. At the same time, recent developments support the hypothesis that ECE countries – in contrast to most nations undergoing regime changes in recent years – will have more difficulties selecting and settling into an "appropriate" type of democracy (Schmitter and Karl 1992, 53).

Comparative analyses of transition problems in post-authoritarian and post-communist societies emphasize four aspects of the consolidation of new democracies (Crawford and Lijphart 1995; Merkel 1994; Sandschneider 1995):

a) The *economic performance approach*. In their classic study of "civic culture", Almond and Verba already pointed to positive economic development as an extraordinarily important factor in the consolidation of post-authoritarian regimes. In their view, efficacy and material output reinforce the legitimacy of democratic systems, while disappointed economic expectations will tend to weaken the legitimacy of new democratic regimes considerably.

b) The *institutional elites approach*. "Whether the new or changed regimes in these countries will be operated by disunified or consensually unified elites" (Highley and Gunther 1992, 344) is equally decisive for the prospects of consolidating new democracies. The institutional elites approach therefore focuses on the capacity for "consensus-building" within political elites as well as on their efficiency and responsiveness, but also on the bonus of confidence the public accords to central institutions and representatives of the political system.

c) The *integration-aggregation approach* deals primarily with the way a democratic system contributes to social integration. Without a minimum of workable intermediary structures of integration and aggregation of social interests (i.e. unions, employers' and employees' associations), new democracies face considerable problems of legitimacy. This mainly concerns the "floating party systems" within ECE states and parties, which are "politically too 'strong', but socially as organizations too 'weak' and unable to maintain a meaningful social dialogue" (Ágh 1994, 236). As a consequence, populist actors are able to enter the "intermediary vacuum" (Plasser and Ulram 1993) with their messages and direct protest appeals, and will exploit existing deficits of representation for their purposes.

d) The *political culture approach.* In keeping with the democratization of post-communist societies, comparative political culture research is undergoing a spectacular revival (Gerlich, Plasser and Ulram 1992; Kaase 1994 a, b; Meyer 1993; Weil 1994). International research networks are competing for the latest data, and the time span between collecting data and publishing research reports is narrowing continuously. Trying to keep up with the dynamics of political change in Eastern Europe increases competition among political scientists. It also raises the danger of superficial or unsubstantiated interpretation. Thus, the lack of trend data as well as conceptual and methodical weaknesses of survey research lead to contradictory statements and over-interpretation of short – lived but suggestive data. Undoubtedly, political culture research in Eastern Europe is still miles away from forging a theory of political culture applicable to these countries (Beyme 1994, 349).

The most frequently voiced – and often justified – criticisms of current practices in comparative political culture research on Eastern Europe include:

– Criticism of a *safari approach* to data gathering (Kaase 1994). Including as many countries and cases as possible into an analysis tends to create a problematic distance between researcher and object of research. It also encourages sterile research designs as well as, more often than not, schematic and inadequately contextualized interpretations.

– Critique of *numbering and ranking.* Insufficient linguistic competence, cultural remoteness, and a poor understanding of current processes and developments in the countries under scrutiny favour mechanical interpretations. The core and key question of political culture research – how individual attitudes on the micro-level can be condensed into parameters of political culture on the macro-level – is then answered by following the simple rule of aggregation implied in the received political culture approach: "The political culture becomes the frequency of different kinds of cognitive, affective and evaluative orientations toward the political system in general, its input and output aspects, and the self as political actor" (Almond and Verba 1963, 17). But this rule is – as Müller (1993, 8) has observed – just as likable as it is unrealistic; and this is especially true wherever the concept of political culture is devised as a variable explaining the status of political systems.

– Finally, a third and more fundamental type of criticism questions the use of mass survey data for the analysis of political culture. Survey data and standardized questionaires – so the argument goes – can do no more than gather short-lived ad hoc opinions and moods. Searching for deeper insights into the dynamics of change in

4

political cultures, these critics argue for qualitative research methods more sensitive to the political culture of daily life than standardized survey instruments.

Despite massive and largely justified criticism, however, mass survey data are still indispensable for any analysis of change in political culture. Comparative political culture research *per se* requires survey data, internationally standardized questionnaires, and intensive trend data. Nevertheless, the professionalization of internationally comparable research, of data surveys and techniques of analysis is confronted with a number of theoretical and methodological problems that have so far defied resolution. We would like to, first, raise these issues, and subsequently illustrate them on the basis of current research on political culture in East Central Europe.

1. Comparative political culture research on East Central Europe is confronted with questions that cannot be answered satisfactorily using established methods of analysis. Primarily, this concerns the issue of democratic consolidation and its progress, of the emergence of stability and resilience in new democracies. If we take consolidation to be a "process by which democracy becomes so broadly and profoundly legitimate among its citizens that it is very unlikely to break down" (Diamond 1994, 15), then the present data allow for speculative conclusions only. Whether ECE countries have already passed the magic stage of consolidation and have evolved from "insecure" to "secure" democracies cannot definitely be answered at the present time. All that political culture research can do now is to make cautious statements concerning the spread of democratic orientations and values. How strong the hold of the new "rules of the game" would be in case of a crisis, and to what extent the public would resist populist appeals and authoritarian solutions are, however, questions that can only be answered tentatively from today's point of view.

2. This brings us to a central problem of comparative political culture research: the difficulty of establishing criteria for measuring the degree and intensity of democratic attitudes within a given population. Is there, for instance, any point in comparing the latest data on ECE countries with empirical findings on the redemocratization phases of post-autoritarian democracies like Austria, Germany or Italy? Apart from the fact that the available data are quite unsatisfactory, it seems clear that the contexts of both waves of democratization differ so fundamentally that trends of the 1950s cannot serve as criteria for evaluating current trajectories in ECE. The same applies to Southern European democratization uring the 1970s. The historical uniqueness of transformations occuring simultaneously in the political and the economic sphere would seem to preclude any linear transfer of trends observed at different times, and under incomparable circumstances.

3. Comparative research on political culture in Eastern Europe suffers from a *tabula rasa* problem: the first – and only partially comparable – data were collected in 1990. Data on patterns of political culture prior to the regime changes are not only very rare, their questions also routinely encourage a heavy pro-regime bias. They allow at best for vague insights into everyday political culture during the communist period. Especially with regard to conspicuous and country-specific differences in central political orientations, varieties of political socialization under communist regimes need to be evaluated. We need to know, in other words, to what extent political experience, values and social expectations of the past are helping to shape present

5

attitudes of democratic actors and institutions. Such retrospective questions about past political experiences in present surveys are, to be sure, no replacement for data on the political culture of the 1970s and 1980s.

4. This *tabula rasa* problem not only concerns the lack of comparative gauges and attitudinal data on the political culture under communist regimes. The absence of close and comparable trend surveys also overshadows current research endeavours. "A mature research of political culture will require longer periods of observation and the results of the presently available data will have to undergo further statistical analyses before a methodical level of civic culture, of political action or of post materialism research can be reached" (Beyme 1994, 332). But the high cost of continuous and comparative trend research can hardly be met by tightly limited ressources and research budgets. The apparent "normalization" of politics in reform countries makes it increasingly difficult to finance complex research projects. For political culture research, this means that available ressources must be used as efficiently as possible, and that much closer cooperation between international research groups should be targeted.[1]

5. In its analyses, political culture research frequently uses the classic approaches of the Almond and Verba-model, which differentiates between input, output and system orientations. Although Almond and Verba introduced the concept of civic culture as a critical gauge in the description and evaluation of Western democracies, this notion somewhat lacks in precision. Even after decades of political stability, economic growth and high standards of welfare, the criteria of civic culture remain to be fulfilled by all of the European Union's core countries except Denmark (Gabriel 1994, 127–131). This inevitably raises the question to what extent established notions of political culture research can or should be applied to new ECE democracies. Typological concepts of subject culture versus civic culture, of state-oriented versus civil-oriented cultures, of action-stimulating versus action-programming cultures etc. are of only limited use even in countries with strong democratic traditions. It seems all the more inadequate to apply them to the new Eastern European democracies without profound adaptation.

6. Where could we look for more realistic approaches? Attempts to explain conspicuous tensions and contradictions within the political cultures of ECE countries indicate three central cleavages which coincide with attitudes and expectations specific to the new democratic system. These are:

a) the tension between potential winners and potential losers of the transformation process;

b) the tension between supporters of a liberal market and competitive society on the one hand, and persons with paternalist attitudes toward the state on the other;

c) the tension between values of self-reliance (civic orientations) and the need for guidance (traditional subject orientations).

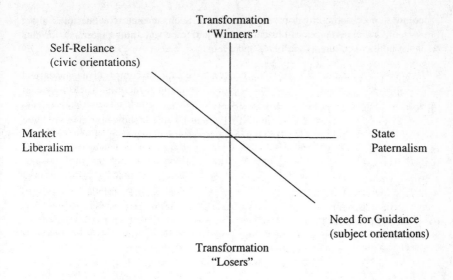

Figure 1 Model of underlying tensions in the political culture of Eastern Europe

While positive individual experience during the transformation process favours active and civic orientations (self-reliance, political efficacy and competence), existential economic problems and the fear of a decline in social status reinforce paternalistic expectations as well as the need for protection and guidance.

In contrast to the comparatively homogenous political cultures of Western European democracies, ECE countries and their political culture appear to be much more fragmented at this point. There is a deep chasm between the political attitudes of the "new entrepreneurial class" – of younger, confident, and upwardly mobile groups – on the one hand, and the paternalist mindset of economically and socially distressed segments of the population on the other. If we are, therefore, to understand the dynamics of change in political culture, it will be necessary to distinguish specific social and political realities within post-communist societies. More qualitative research – in addition to large scale mass survey data – is indispensable for a better understanding of the dynamics and counter-dynamics of changes in political culture.

Economic development and economic orientations

The new democracies of East Central Europe are industrialized countries, though by no means wealthy ones. Using GDP per capita on exchange rate basis as an indicator, the Polish and Slovak figures amount to roughly one tenth of the American, German and Austrian equivalents, while Czech and Hungarian figures are slightly higher. Comparisons on exchange rate basis, however, tend to underrate the economic strength of countries with a large non-market and "grey" economic sector. Using GDP per capita according to purchasing power parities, the Czech figures amount to approximately one third of the Austrian, or one fourth of the American ones; Hungary and the Slovak Republic come next, while the Polish figure amounts to about one fourth of its Austrian or one fifth of its American equivalent.[2]

The disastrous economic situation inherited from the previous regimes, the aftermaths of COMECON's economic dissolution, and particularly the near collapse of trade with the former Soviet Union, exacerbated problems of transferring to a market economy and caused serious syndromes of crisis in all ECE countries. In Poland, and subsequently in Hungary and in the ČSFR, GNP and purchasing power declined sharply, while inflation rates increased drastically. In addition, the reform states were suddenly struck by the hitherto unknown phenomenon of unemployment. Their social security systems, however, were neither prepared nor able to draw on adequate financial resources to deal with these new challenges. Since the first steps of economic transformation coincided with a global recession, the possibilities of capital transfer to the East was as limited as the scope for export of East Central European goods and (legal) labour. The economic reforms are, on the other hand, beginning to show positive results. Poland was the first to suffer, but also the first to recover with a sizable GDP growth in 1993 and 1994. The Czech Republic also shows a continuous positive trend. Hungary and Slovakia registered a considerable GDP growth in 1994, but both are preparing for a slowdown in the near future. Inflation and unemployment will remain high in most countries (except in the Czech Republic), and real wages will generally be low by Western standards.[3]

Table 1 Economic indicators: East Central Europe (1989–1995)

Change in percent	1989	1990	1991	1992	1993	1994 estimate	1995 forecast
Poland							
GDP	+0.2	−11.6	−7.0	+2.6	+3.8	+5.1	+6.0
real wages[1]	+12	−25	−0.1	−2.7	−2.9	+1.7	+3.0
consumer prices	+251	+586	+70	+43	+35	+32	+29
unemployment rate[3]		6.3	11.8	13.6	16.4	16.0	16.0
Hungary							
GDP	−2.0	−3.5	−11.9	−3.0	−0.8	+2.0	+1.0
real wages[2]	+1	−4	−7	−1	−4	+7	−8
consumer prices	+17	+29	+35	+23	+23	+19	+29
unemployment rate[3]	7.8	13.2	12.1	12.0	12.6	10.9	11.0
Slovak Republic							
GDP			−14.5	−7.0	−4.1	+4.8	+4.0
real wages[1]	+1	−6	−25	+9	−4	+3	+3
consumer prices	+1	+11	+61	+10	+23	+12	+12
unemployment rate[3]		1.6	11.8	10.4	14.4	14.8	15
Czech Republic							
GDP	+1.5	−2.0	−14.2	−6.4	−0.9	+2.6	+4.0
real wages[1]	+1	−5	−24	+10	+4	+7	+8
consumer prices	+1	+10	+57	+11	+21	+10	+10
unemployment rate[3]		0.6	4.1	2.6	3.5	3.5	3.5

1) average monthly wages
2) average gross wages
3) unemployment rate: percentage *not* change

Source: Österreichisches Institut für Wirtschaftsforschung – WIFO Monatsberichte 1993–95, OECD, EIU. Wiener Institut für Wirtschaftsvergleiche

Since 1992, and especially so in Poland and the Czech Republic, popular evaluation of the overall situation has improved gradually. Fear of unemployment is also declining, although a majority of respondents in all East Central European countries still think that it will rise in the near future.

In 1994, about two thirds of all respondents in Hungary and Poland, and more than half in Slovakia, considered their regular incomes as near or on the poverty line, and even more describe the economic situation of their families as unsatisfactory. Only in the Czech Republic more than 50% consider their regular incomes and their economic situation as sufficient or satisfactory, respectively. The prevailing attitude of respondents is that their financial situation is now worse than prior to the regime changes–only in the Czech Republic roughly 50% of respondents note an improvement or at least a stabilization of their financial situation.

These evaluations of economic conditions clearly influence the perception of the new political system's performance. In all countries, those who expect negative economic developments, further sharp increases of inflation and shrinking incomes declare dissatisfaction with the political system, while those with an optimistic or neutral outlook ("things will remain much the same") tend to show a positive view of the new system.[4]

Table 2 Evaluations of general economic and employment situation – East Central Europe (1992 and 1994)

in percent	Hungary			Poland			Slovak Republic			Czech Republic		
	92	93	94	92	93	94	92	93	94	92	93	94
This year's economic situation is												
– much/somewhat better than 12 months ago	11	8	12	8	16	22	9	5		19	24	26
– about the same	12	18	25	10	19	24	11	14		18	21	29
– somewhat worse	35	35	30	26	27	24	35	45		35	34	27
– much worse than 12 months ago	39	35	28	52	32	19	36	29		21	16	11
Unemployment in the next 12 months will be												
– substantially/somewhat lower than this year	0	7	18	4	7	12	6	5		4	6	7
– about the same	3	13	25	6	9	19	5	11		11	15	28
– somewhat higher	32	46	35	34	39	35	38	47		46	49	42
– much higher than this year	59	28	12	49	38	20	42	26		34	21	13

Source: Fessel+GfK, Economic Surveys in East-Central Europe (1992, 1993, 1994)

**Table 3 Evaluations of subjective economic situation: East Central
Europe (1991–1994)**

in percent of a) total population b) regularly employed	Hungary			Poland		
	1991	1992	1994	1991	1992	1994
b) To make ends meet, regular income is						
− definitely/ rather sufficient	26	26	29	39	34	36
− not really sufficient	56	54	39	37	39	41
− definitely insufficient	18	19	28	24	26	21
b) All income sources are sufficient to earn a living		64			62	

a) The present economic
   situation of the family is
   seen as						

− highly/rather satisfactory	24	20	19	18	23	24
− not very satisfactory	50	47	48	56	49	52
− highly unsatisfactory	26	33	32	26	26	23

a) In the last five years the
   financial situation of the
   household has						

− much/somewhat improved	10	7	8	18	14	17
− remained the same	23	18	18	22	18	18
− worsened a little	45	47	42	37	35	32
− much worsened	22	27	33	23	31	31

Table 3 (continued)

in percent of a) total population b) regularely employed	Slovak Republic				Czech Republic			
	1991	1992	1993	1994	1991	1992	1993	1994
b) To make ends meet, regular income is								
– definitely/ rather sufficient	39	37	35	46	46	51	51	56
– not really sufficient	33	44	44	35	35	33	36	32
– definitely insufficient	28	19	20	17	18	14	12	11
b) All income sources are sufficient to earn a living			65				72	
a) The present economic situation of the family is seen as								
– highly/rather satisfactory	48	49	48	42	59	60	60	53
– not very satisfactory	42	43	40	41	34	34	35	39
– highly unsatisfactory	10	8	11	17	7	5	5	8
a) In the last five years the financial situation of the household has								
– much/somewhat improved	15	21	17	19	18	22	22	25
– remained the same	20	19	18	21	29	25	28	26
– worsened a little	37	37	40	35	33	30	32	30
– much worsened	30	22	24	26	20	21	18	19

Source: Fessel+GfK and Paul Lazarsfeld-Gesellschaft, NDB-Neue Demokratien Barometer (1991–1994)

Even though the performance of the present economic system is viewed rather critically and there are signs of nostalgia for the old ways (Rose and Haerpfer 1993; Plasser and Ulram 1993b) – where most people "had little but the little seemed to be guaranteed" – there is no real urge to return to a planned economy. In all East Central European countries a vast majority generally welcomes the economic reform process. Even in Russia, privatization is viewed favourably, albeit to a lesser extent.[5] For the future, rays of hope are thus brightening the horizon. In all countries, most notably in the Czech Republic, a (relative) majority of respondents expect their personal economic situation to improve within the next five years, and the number of those who fear ongoing income losses is decreasing.

Relative improvement does not, however, mean satisfaction: especially in Poland and Hungary, only a small minority is convinced that a satisfactory living standard will be reached in the foreseeable future. Furthermore, evaluation of living standards not only reflects "objective conditions" but is also influenced by the "subjective factor of rising expectations". One should note that, while the level of satisfaction with (the possession of) advanced consumer goods has declined all over East Central Europe during the last year, actual household ownership of exactly the same type of goods has increased considerably and old products have been substituted by new ones, often enough of Western provenance.[6] Evidently people now tend to see their individual situation in the light of a seemingly comparable one "on their very doorsteps" in Western Europe and/or in that of the amount and quality of goods available in their new market economies.

**Table 4 Prospective evaluation of subjective economic situation
(1991–1994)**

in percent of total population	year	Economic situation of household in five years will be		
		much/ a little better	about the same	worse than today
Hungary	1991	48	28	24
	1992	43	30	24
	1994	43	33	20
Poland	1991	42	31	27
	1992	32	30	31
	1994	42	33	15
Slovak Republic	1991	47	27	15
	1992	53	22	24
	1993	49	22	27
	1994	49	26	24
Czech Republic	1991	60	24	16
	1992	55	26	17
	1993	57	26	15
	1994	55	30	14

Table 4 (continued)

in percent of total population	year	Personal living standard will reach a satisfying level				
		within the next 5 years or is already adequate	in 6 to 10 years	in more than 10 years	never	undecided
Hungary	1991	20	12	10	18	36
	1992	23	17	13	17	30
	1994	17	14	14	21	33
Poland	1991	14	10	11	22	42
	1992	19	8	9	25	47
	1994	19	8	13	24	43
Slovak Republic	1991	27	21	18	6	29
	1992	35	29	18	10	8
	1993	37	18	14	18	12
	1994	21	16	18	7	38
Czech Republic	1991	32	21	17	6	24
	1992	33	25	21	13	7
	1993	34	23	20	13	9
	1994	25	17	12	8	37

Source: Fessel+GfK and Paul-Lazarsfeld-Gesellschaft, NDB-Neue Demokratien Barometer (1991–1994)

In the political arena, general approval of economic reforms and the market economy can be, and – as shown by election results in Poland and Hungary – often is compatible with strong criticism of particular aspects of the reform process (social hardship, economic scandals etc.). Only in exceptional cases do East Central Europeans cherish the ideal of an (almost) fully liberalized economy. In all four countries, the majority favours a system in which the state reaches key economic decisions in cooperation with large interest groups (trade unions, employers associations), i.e. some form of social partnership or neocorporatism along the lines of those existing in Austria or Sweden. Only minorities give preference to either a liberal economy with a minimum of governmental interference – in which the state essentially limits itself to setting up a framework for economic activities, e.g. tax laws and procedural rules of competition – or favour a planned economy.

Table 5 State and economy in East Central Europe (1993/94)

in percent prefer	Hungary		Poland		Slovak Republic		Czech Republic	
	1993	1994	1993	1994	1993	1994	1993	1994
minimal state	19	28	13	19	18	24	22	26
neocorporatism	55	57	56	61	50	53	57	51
state planned economy	20	12	27	16	30	22	19	21
no response	6	3	4	4	2	1	2	1

Source: Fessel+GfK, Vertrauen in Finanzinstitutionen in Ost-Mitteleuropa (1993) and Fessel+GFK, Politische Kultur in Ost-Mitteleuropa (1994).

The desired social partnerships or neocorporatist dealings between state and economy depend, however, on a number of structural, institutional, and political-cultural preconditions. These normally do not exist in adequate form in the new democracies of East Central Europe, nor could sheer voluntarism produce them. Of these prerequisites, the following should be mentioned in particular: strong interest groups both in terms of membership and of determination, which also enjoy a large basis of legitimacy; well-functioning government institutions; considerable national autonomy in economic decision making; and a pragmatic and consensual style of policy formulation and interest representation.[7]

Confidence in ruling elites and political institutions

Institutional confidence profiles in the new Eastern European democracies show distinct country-specific variations. In the Czech and Slovak Republics as well as in Hungary, presidents enjoy the highest level of trust. In these parliamentary systems, presidents evidently benefit from an image of statesmanship, and from the understanding that they are moral guardians placed above everyday political struggle. They are widely seen as aiming at consensus, balancing interests and ensuring strict observation of constitutional rules. In countries whose presidents also exercise executive functions and either formulate or at least strongly influence government policies, the situation is entirely different. In Poland, the president wields strong influence and enjoys extensive veto rights. Similarly, the constitution of the Russian Republic allocates wide ranging executive and directive powers to the president. Both Walesa and Jelzin, however, suffer from severe confidence gaps. Both countries lack a political and moral authority untarnished by the process of political competition and serving as "presidential-charismatic guardian of the constitution" (Offe 1994).

In Russia, the institutional confidence gap is especially marked. Except for the army, which seems to have arrogated the task of guarding national or patriotic unity, all central political institutions meet with pervasive distrust and scepticism. Especially disquieting is not only the scant confidence in government and law courts, but also the prevailing distrust vis-à-vis the elected parliament. In Poland, too, police and army have evidently gained credibility, while political elites have lost it. Traditional institutions like the church and law courts are backed by higher levels of public approval than the democratically elected parliament or the government in office. A similar pattern can be found in Hungary and in the Slovak Republic, except for a satisfying level of trust enjoyed by the presidents of both countries. But even here, church(es) and army are clearly considered more trustworthy than government and parliament. Signs of a confidence gap are also clouding the political horizon in East Germany, where institutional confidence is decidedly lower than in the West.

In the Czech Republic, the profile of institutional confidence looks markedly different. Here, both president and government enjoy a high level of credibility. Media and law courts appear more trustworthy than army or churches. Confidence in parliament is comparatively low, though still higher than in other East Central European countries.

Table 6 Confidence in institutions in East Central Europe (1994)

in percent	East Germany	Czech Republic	Slovak Republic	Hungary	Poland	Russian Republic
president	–	68	61	65	19	24
army	39	39	47	44	59	50
police	36	36	31	45	41	21
courts	36	40	33	46	34	25
public administration	22	27	26	33	24	18
government	25	56	31	21	25	19
parliament	26	32	22	23	24	12
churches	24	28	45	39	36	–
mass media	20	43	38	30	34	30
old labour unions (close to CP)	–	11	12	14	13	–
new democratic unions	41	20	22	23	14	14
political parties	18	24	16	11	8	14

Note: Values 5–7 on a 7–grade scale from 1,0 (= no confidence) to 7,0 (= strong confidence)

Source: Fessel+GfK and Paul Lazarsfeld-Gesellschaft, NDB-Neue Demokratien Barometer (1994) Gluchowski and Zelle (1993) for East-Germany (1993), GfK-Moscow, Social-Political Survey (1994)

A deeper insight into the dynamics of institutional confidence can be gained by trend analyses (Plasser and Ulram 1994a), which show distinct patterns of development in each country. In Hungary, practically all institutions were affected by a decline of confidence between 1989 and 1992, a trend only recently reversed with respect to state institutions such as law courts, police and armed forces. On the other hand, the downturn continues for churches and media. Parliament and political parties still do not inspire confidence, which also holds true for unions. New democratic unions are now gaining in confidence while old ones (those associated with the former Communist Party) are losing it. In Poland, most state institutions have managed to consolidate their position. The credibility of parliament has increased in the wake of the 1993 elections but is still low, as is that of the government. Both the Catholic Church and the Solidarity trade union have been suffering from an ongoing crisis of confidence. Thus, the formerly negative image of police and armed forces, which were discredited as a result of their association with the old regime and their involvement in the imposition of martial law, has now given way to a more "neutral" standing as a protective and law enforcing function. Those institutions which originally supported democratic transition (Church, Solidarity) have evidently exhausted their confidence bonus. The same holds true for president Walesa, who is the only head of state in the new East Central European democracies with a negative balance of trust, and scoring only little higher than parties and unions. Slovakia has gone through two stages: initially, toward the end of 1992, autonomous statehood led to greater confidence in "one's own" nation-state institutions, especially in those of state and government. This upward trend, however, did not last long. The new government quickly lost its advance credibility; parliament, parties and unions at all failed to capitalize on the momentum of independence. Only in the Czech Republic is the situation relatively stable. Here, the level of confidence – in political institutions generally, but most of all in the government – is markedly higher than in all other ECE countries.

Table 7 Confidence in institutions in international comparison

in percent	East Germany		Poland					Hungary
	1991	1993	1991	1992	1994	1991	1992	1994
president	–	–	–	–	19	–	–	19
army	35	39	53	63	59	47	40	44
police	37	36	30	46	41	43	34	45
law courts	44	36	30	32	34	44	40	46
public administration	24	22	19	23	24	33	24	33
government		25	26	22	25	26	16	21
parliament	36	26	18	12	24	29	20	23
churches	–	24	50	41	36	46	42	39
media	17	20	30	35	34	40	34	30
old labor unions (close to PC)	39	–	18	21	13	20	20	14
new democratic unions	39	41	20	9	14	25	16	23
political parties	19	18	7	8	8	16	10	11

Note: Values 5–7 on a 7–grade scale from 1,0 (= no confidence) to 7,0 (= strong confidence)

Table 7 (continued)

in percent	Slovak Republic				Czech Republic				Russian Republic
	1991	1992	1993	1994	1991	1992	1993	1994	1994
president	–	–	–	68	–	–	–	68	24
army	31	52	44	47	52	42	40	39	50
police	35	35	31	31	33	36	36	36	21
law courts	39	37	29	33	39	38	32	40	25
public administration	28	30	26	26	25	27	24	27	18
government	23	49	33	31	50	48	48	56	19
parliament	34	24	20	22	32	19	25	32	12
churches	21	45	39	45	27	33	27	28	–
media	47	39	39	38	42	40	41	43	30
old labor unions (close to CP)	39	31	21	12	35	24	21	11	–
new democratic unions	39	31	21	22	35	24	21	20	14
political parties	20	20	12	15	22	20	20	24	14

Note: Values 5–7 on a 7–grade scale from 1,0 (= no confidence) to 7,0 (= strong confidence)

Source: Fessel+GfK. Demokratie- und Parlamentsverständnis in Ost-Mitteleuropa (1991), Fessel+GfK, Politische Kultur in Ost-Mitteleuropa (1992), Fessel+GfK, Politische Kultur in der CSFR/II (1993), Fessel+GfK and Paul Lazarsfeld-Gesellschaft, NDB-Neue Demokratien Barometer (1994), Gluchowski and Zelle (1993) for East-Germany, GfK-Moscow, Social-Political Survey (1994)

Deficits of representation in the new democracies

Institutional deficits of confidence, as well as rampant criticism of political elites and parliamentary representatives, are closely connected with the party system's considerable weakness of integration (Olson 1994). Only a minority of citizens and voters strongly support a political party. Stable party identifications and effective loyalties can be found only rudimentarily, and a majority of the population views parties from a sceptical distance. The "typical" voter in the new Eastern European democracies is a "non aligned voter". Sudden and dramatic changes within the political spectrum, an exceptionally high volatility and fluctuating emotions – as they became visible at recent elections – within the electorate are thus also a consequence of weak parties, of a party system lacking in contours, and of deficient structures of integration.

Table 8 Level of party identification in comparison

percent of party identification	1991	1992	1993	1994
West Germany	60	–	63	–
East Germany	70	–	48	–
Austria	49	–	–	44
Czech Republic	31	37	36	38
Slovak Republic	25	35	31	30
Russian Republic	–	–	–	25
Hungary	25	23	–	22
Poland	15	13	–	15

Source: Fessel+GfK, *Political Indicators* (1991–1994), Plasser and Ulram (1993) for East Central Europe. Gluchowski and Zelle (1993) for FRG, GfK-Moscow, Social-Political Survey (1994)

While East Germany showed – in reaction to economic malaise and a structural "crisis of unification" – a drastic decline in party identification, "party building" and attempts at integration (of, in the main, ruling ODS) through political organization have at least partially been successful in the Czech Republic. There, the degree of party identification (macro-partisanship) is approaching current levels in Austria (where, however, it has been declining sharply since the mid–1980s). The situation remains largely unchanged, on the other hand, in Poland and Hungary. Both party systems suffer from substantial deficits of integration. Only one in six persons identifies with a particular political party. Drastic shifts between parliamentary forces and conspicuous electoral successes of (former) communist parties (in Hungary and Poland) are therefore primarily indicative of diffuse emotions and attitudes of general resentment. These electoral successes are neither grounded in newly formed coalitions of voters and parties, nor in a "realignment" of the (old) party systems, but are the result of highly volatile emotions and beliefs which have been fed by disappointed expectations. This holds true also for the current situation in the Russian Republic, where three quarters of the electorate have no recognizeable ties to any political party.

The organizational and structural weaknesses of party systems in post-communist democracies of Eastern Europe not only favour electoral instability, shifting majorities and fragile coalitions. They also reinforce a critical distance on the part of the electorate toward political elites and central democratic institutions. An "intermediary vacuum" and "floating" party systems are as characteristic of politics in Eastern Europe as they generally are of the societies in these countries. This not only hampers the integration of the electorate into – however rudimentary and unstable – political structures; it also favours the rise of populist leaders. Widespread economic malaise, as well as criticism of the performance of institutional elites and representatives, present many opportunities for mobilization by populist movements of a nationalist or chauvinist character.

In the face of global economic recession which is also slowing down the progress of transformation to a market economy in Eastern Europe, political dissatisfaction has increased markedly, especially in the "crisis year" of 1993. Except for the Czech Republic, where economic and political consolidation has progressed farther than elsewhere, recessions and economic crises have led to a dramatic loss of confidence and substantial decline of approval with the political system. More than half of all respondents in Poland, Hungary and Slovakia declared themselves dissatisfied with democracy and the entire political system of their country at the end of 1993.[8] In 1994, however, the data show a change of heart in the Slovak Republic, in Hungary and, though to a lesser extent, in Poland.

Table 9 Political satisfaction in Eastern and Western countries of Central Europe

		Satisfaction with democracy and the entire political system in percent		
	year	very satisfied	rather satisfied	not satisfied
Poland	1991	5	62	32
	1992	1	54	39
	1993	2	35	57
	1994	2	43	53
Hungary	1991	2	56	39
	1992	3	55	40
	1993	1	33	64
	1994	1	47	48
Slovak Republic	1990	9	55	28
	1991	2	6	41
	1992	1	68	29
	1993/1	2	61	36
	1993/2	2	29	67
	1994	3	70	27
Czech Republic	1990	14	60	19
	1991	4	72	24
	1992	4	69	26
	1993/1	6	73	20
	1993/2	4	60	35
	1994	3	59	37
East Germany	1991	5	62	32
	1993	2	51	41
West Germany	1984	32	61	6
	1985	37	54	7
	1988	24	66	10
	1989	21	68	7
	1991	18	63	12
	1993	12	59	25
Austria	1984	15	69	13
	1985	9	71	18
	1988	9	72	17
	1989	9	73	17
	1991	15	75	9
	1993	5	71	23
	1995	7	59	32

Sources: Fessel+GfK, Politische Indikatoren (1984–1993) for Austria, Fessel+GfK, Vertrauen in Finanzsituationen für PL, H 93 und CZ, SK 93/2, Plasser and Ulram (1993), Fessel+GfK, Politische Kultur in Ost-Mitteleuropa (1994/2)

Close analysis of the factors which stimulate a high level of political discontentment clearly indicates three groups of factors, all of which can be found in the new democracies to a varying degree: economic factors that correlate with a substantial decline in material living conditions and living standards, with the spectre of unemployment and the fear of losing social status; political performance factors, which express criticism of politicians and their actions, of intractable problems and insufficiently responsive political elites, or which manifest themselves in more diffuse reactions of weariness; and finally, antagonistic factors stemming from proximity to parties of the old regimes, from categorical doubts concerning the market economy, from authoritarian attitudes, and from social integration within the webs of "old" structures and ways of thinking (Plasser and Ulram 1994a).

Table 10 Dissatisfaction with democracy in post-authoritarian and post-communist systems during the first stage of consolidation

dissatisfied in %	Southern Europe			East Central Europe					
	1978	1985	1989	1990	1991	1992	1993/1	1993/2	1994
Spain	42	38	36	–	–	–	–	–	–
Portugal	–	53	35	–	–	–	–	–	–
Greece	–	42	42	–	–	–	–	–	–
Czech Republic	–	–	–	19	24	26	20	35	37
Slovak Republic	–	–	–	28	41	29	36	67	27
Hungary	–	–	–	–	39	40	–	64	48
Poland	–	–	–	–	32	39	–	57	53

Note:
1) Question wording for Southern Europe: "On the whole, are you very satisfied, fairly satisfied, not very satisfied, or not at all satisfied with the way democracy works" (in your country)? Percentages of "not very" and "not at all satisfied" respectively.
2) Question wording for East-Central Europe: "On the whole, are you very satisfied, fairly satisfied or not satisfied with democracy (in your country)? Percentages of "not satisfied".

Source: Montero and Torcal (1990) for Southern Europe, Plasser and Ulram (1993) for East Central Europe, Fessel+GfK, Politische Kultur in Ost-Mitteleuropa (1994)

Apart from the rise of political dissatisfaction during the second half of 1993, which was primarily caused by the recession, these results begin to look less dramatic when compared with levels of approval with the political system found during the "second" transition stage of Southern European democracies. In Spain, Portugal or Greece, the emerging consolidation was similarly marked by considerable ups and downs in the satisfaction with the political system. Fluctuations in the economy or internal situative conflicts could substantially alter levels of political approval without ever questioning the progress of democratization as such. Furthermore, comparable data on satisfaction with the workings of democracy in Western Europe in 1994 show results similar to those found for East Central Europe in 1993. According to a Eurobarometer survey of

1994, 50% of EU citizens are unhappy with democratic procedures in their country (66% in Greece, 67% in Spain, 70% in Italy). The indicator of "democracy satisfaction" obviously reveals mainly "specific support", and includes aspects such as the evaluation of economic performance or the political behavior and public image of ruling elites.

Political culture and democratic consciousness

In contrast to "hard" economic conditions of consolidation in post communist democracies, factors of political culture are part of the "soft sphere" (Ágh 1993) of system transformation. The landscape of political culture in the new democracies is marked by tensions between, on the one hand, traditionalist and parochial orientations linked to experiences of political socialization during decades of communist rule, and, on the other hand, nascent democratic and pluralist values. This raises the question, to what extent multi-party systems and party competition are accepted as rules of the game and basic constituents of a democratic order. In the Czech Republic, there is hardly any opposition to a multi-party system: the dominant attitude is comparable to the one prevalent in Austria. 20% of respondents in Slovakia, 22% in Hungary and 23% in Poland do, however, prefer a single-party system over one of competing parties. Preference for a single-party system in Hungary and Slovakia has actually increased in recent years. In Poland, aversion to party competition first rose sharply from 19% (1991) to 31% (1992), but subsequently declined after parliamentary elections in 1993.

Table 11 Preference for a single-party system in post-authoritarian and post-communist countries

Preference for a single-party-system (in %)	early 50s	late 50s	late 70s	1991	1992	1993	1994
West Germany	21	12	8	–	–	–	–
Austria	–	16	11	5	–	–	–
Italy	–	–	6	–	–	–	–
Czech Republic	–	–	–	6	8	8	6
Slovak Republic	–	–	–	14	14	16	20
Hungary	–	–	–	18	22	–	22
Poland	–	–	–	19	31	–	23

Source: Plasser and Ulram (1993b), Fessel+GfK, Paul Lazarsfeld-Gesellschaft and NDB-Neue Demokratien Barometer (1994).

Comparative data on the consolidation stages of German, Italian and Austrian post-war democracies show that anti-pluralist sentiments survived in these post-authoritarian democracies, too. But they gradually subsided as the new political and economic institutions proved their ability to function. Despite considerable economic problems and widespread political discontent, an overwhelming majority of citizens in East Central Europe clearly prefer democracy to authoritarian and dictatorial regimes. Even in the Russian Republic, where pre-democratic and anti-pluralist attitudes are more persistent, the first free parliamentary elections – as well as the shock of an attempted coup d'état carried out by adherents of the old regime – have helped to strengthen democratic commitment. Supporters of an authoritarian regime have, however, also reconstituted themselves; the choice between democratic and authoritarian types of leadership has therefore yet to be made a definite one in the Russian Republic – quite contrary to reform countries of East Central Europe.

Critical views on the current performance of political and economic institutions do not necessarily entail outright rejection of democracy. After all, between two thirds (in Poland and the Slovak Republic) and three quarters (in Hungary and the Czech Republic) of respondents express a basic preference for a democratic order over any authoritarian variety. Anti-democratic preferences, in combination with authoritarian-populist sentiments, certainly do provide a fertile ground for any potential of political destabilization. This potential, however, has not yet acquired a momentum sufficient to threaten the systemic stability of East Central Europe's new democracies. Though scarce, empirical evidence for early postwar Austria and (West) Germany shows clearly that the early stages of democratization were equally overshadowed by antidemocratic resentments and orientations. With time, such predilections gradually ebbed away as a consequence of positive economic and political performance of the new market-orientated pluralistic democracies, and of the integrative capacity of political and societal actors, that is, of organized interests. The situation looks rather different in the Russian Republic. There, anti-pluralist and authoritarian tendencies, as well as nationalist and populist megalomania, are backed by a relatively large sector of the public.

Table 12 Democratic consciousness in post-communist countries over time

in percent agreeing	year	democracy preferable in any case	dictatorship possibly better	makes no difference to people like me	no response do not know
Czech Republic	1990	72	8	12	9
	1991	77	7	15	1
	1992	71	10	18	1
	1993	72	9	17	1
	1994	75	11	14	1
Slovak Republic	1990	63	11	18	8
	1991	67	10	22	1
	1992	68	11	19	2
	1993	60	11	28	1
	1994	68	11	19	2
Poland	1991	60	14	23	3
	1992	48	16	30	6
	1994	64	17	16	3
Hungary	1991	69	9	18	4
	1992	69	8	21	2
	1994	73	8	16	3
Russian Republic	1993	36	20	19	25
	1994	50	27	23	1

Source: Plasser and Ulram (1993), Fessel+GfK, Paul Lazarsfeld-Gesellschaft and NDB-Neue Demokratien, Barometer (1994), GfK-Moskow, Social-Political Surveys (1993–1994)

Some tentative conclusions

Our analysis of the consolidation process in Eastern Europe has shown quite distinct patterns of change in all four countries. Finally, we will try to summarize some of the more important factors that shape the course and the chances for political consolidation in post-communist systems.

Economic performance The breakdown of planned economies and the creation of markets were followed by crises that hit all ECE countries in similar ways. Poland was the first country to adopt a policy of strict liberalization, and Hungary had already taken steps in this direction under the old regime. In both countries, the social costs of "transformation began to be felt at an early stage, since economic change rapidly increased existing income disparities, created new ones and thus produced a considerable stratum of poverty. In this respect, the Czech and Slovak Republics were latecomers – though the former is the industrially more advanced country in a region with less marked internal differences – and thus suffered much less from the break-up of Czechoslovakia. Living standards are highest and (on the whole) comparable in Hungary and the Czech Republic, as is the anticipated economic upswing in the nearer future. But the evaluation of both general and individual economic prospects differs between them, Hungarians expressing a very pessimistic, and Czechs a rather optimistic outlook. Evidently Hungarians, who once considered themselves forerunners of economic reform, react to economic problems with particular sensitivity (Ilonszki and Kurtán 1992; Plasser and Ulram 1993 c and d).

Cultural cleavages and political culture Both the Czech Republic and Hungary are highly secularized countries, though the latter has preserved stronger parochial and paternalist ties, especially in more remote areas. The same is true to an even greater extent of Slovakia and Poland. Democratic traditions – although interrupted by four decades of, first, totalitarian and subsequently authoritarian communist rule – also differ markedly. The Czech Republic clearly has the longest democratic tradition[9], having maintained a democratic system even during the interwar period. In contrast, the democratic traditions of Poland, Hungary and even Slovakia (which was the object of patronizing tutelage on the part of Czechia during the first Czechoslovak Republic) were much shorter and interrupted by spouts of authoritarianism long before German and Russian occupation or domination. In Poland, the Catholic Church played a prominent part in anticommunist resistance, but had visible difficulties in repositioning itself within the new democratic order. It thus contributed to a widening cleavage between churches and the secular state, which cut across other political divisions and intensified ideological conflict. Slovakia's political culture in turn reflects problems of nation-building as well as the difficulties of coming to terms with a strong Hungarian minority. The Czech Republic, on the whole, shows the most homogeneous political culture characterized by rather pragmatic attitudes towards politics, by a higher level of (perceived) political efficacy and confidence (Plasser and Ulram 1993 c and d) and by comparatively stable democratic orientations.

Political structure and institutional differentiation In general, Eastern European political parties are only loosely rooted in social interests and public awareness. Social and cultural cleavages are less pronounced than in Western Europe and are – with some exceptions (Slovakia) – hardly reflected in party systems. Following the disintegration of anti-communist coalitions which had dominated the founding elections between 1989 to 1991, party systems became even more fragmented, internal divisions and splits being on the order of the day and complicating further the political map of competition and responsibility. Only in rare cases (notably those of ODS in the Czech Republic) successful steps were taken to erect stable organizational infrastructures or efficient, checked and balanced institutional frameworks (with the partial exception of Hungary: Ilonszki and Kurtán 1993). This problem has proved especially pervasive in (more or less elaborate) presidential systems such as those of Poland and Russia. The civil service was not only dominated by adherents of the old regime, but often underwent aggressive political penetration and manipulation by new parties. The same is true for the media, and most notably for public television.[10] Confidence in political institutions, political leaders, in the civil service and in the administration of law is correspondingly low. The resulting intermediary vacuum provides fertile ground and a wide range of opportunities for diffuse protest and populist leaders.

Elite behaviour The new political elites have primarily been recruited from democratic opposition movements. They may be characterized by high moral standards and exceptional personal integrity on the one hand, and by an often somewhat missionary approach to politics and lack of experience with ordinary administration or political bargaining on the other hand. In Poland and Hungary, where the transformation process had started early and pluralization advanced more rapidly due to a relatively liberal political climate, these new political elites found themselves confronted with challenges they were not prepared for. Here, as in Slovakia with its focus on the "national question", political discussion among elites tended to focus on values and ideological commitments: This priority, however, did not correspond with the pressing concerns of a majority of the population. Thus, politics neither provided an intelligible and reassuring agenda for the future, nor did it concentrate on pragmatic policy making and consensus building. Only in the Czech Republic, "professional" and "pragmatic" politics took over after a relatively short interlude of "unpolitical politics" (Brokl and Mansfeldova 1992), the dominant political elites acting in near unison with respect to the principal direction political and economic development should follow, and taking care to maintain the dialogue with other societal actors like the media or trade unions.[11]

Different economic and political trends – and their even more diverse perception by the public – as well as elite reactions specific to each country and the complex challenges of transformation processes call for caution against linear or determinist models of democratization in post-communist systems. Although positive economic factors obviously favour the development of positive attitudes towards rules of democratic and market competition, growth rates and positive evaluations of subjective economic conditions can only serve as limited indicators of the prospects for political consolidation.

Table 13 Anti-democratic orientations by subjective economic climate and party identification: East Central Europe (1994)

in percent	total	Satisfied by recent economic situation of the family		Expected financial situation of household in 5 years from now		Respondents have party-identification	
		yes	no	better	worse	yes	no

Preference for a single party system

Poland	23	22	23	23	25	26	24
Hungary	22	16	31	18	36	19	26
Slovak Republic	20	17	25	15	24	20	20
Czech Republic	6	4	13	5	9	3	7

Democracy or dictatorship makes no difference for people like me

Poland	16	10	23	12	20	12	18
Hungary	16	10	26	11	26	6	20
Slovak Republic	19	14	30	17	26	11	21
Czech Republic	14	10	24	11	20	6	19

Dictatorship can be better under certain circumstances

Poland	17	18	15	13	21	14	18
Hungary	8	9	9	8	9	14	6
Slovak Republic	11	10	13	9	14	9	12
Czech Republic	11	10	10	8	20	10	11

Source: Fessel+GfK, Political Culture in East-Central Europe (1994), Fessel+GfK and Paul Lazarsfeld-Gesellschaft, NDB-Neue Demokratien Barometer (1994).

The economic performance approach should thus be supplemented by approaches which take account of such criteria as political performance, efficiency of ruling elites, and the integrative capacities of societal actors like parties and organized interests. Rather than looking for a general theory of democratization, it may prove more rewarding to trace corridors of action in each country, and to examine strategic resources of political as well as societal actors.

Notes

1 Although a separate subdiscipline of "comparative transition research" has been established – which originally focused on so-called Third World countries and Southern European nations undergoing system changes – comparative empirical studies of system transformation in several countries and including subsequent stages of consolidation are few and far between. The more important of them are (in so far as they have been made accessible through publication)
- the Central and Eastern Eurobarometer 1990/91 and 1991/92, (Eurobarometer 1992/93),
- the Times Mirror Study of 1991 (Times Mirror, 1991),
- the USIA-Surveys of reform countries (McIntosh, MacIver and Abele 1992),
- the New Democracy-Barometer (1991, 1992, 1994) of the British-Austrian research group Rose and Haerpfer (1991 and 1993).
- comparative political culture studies for East Central Europe partially complemented by contributions to the political culture of Russia, Slovenia and the former German Democratic Republic by the Austrian research group Gerlich, Plasser and Ulram (1992) and by Plasser and Ulram (1993b).
The first two studies are, geographically speaking, broadly conceived but methodologically and conceptually inadequate. The New Democracy Barometer concentrates primarily on aspects of economic policy and the economy in general. Data regarding political orientations are still underrepresented, but do allow insights into perceptions of system change by those involved in it.
In addition, there are many national studies, some of them attempting to trace (national) developments with considerable methodological sophistication. A major problem of existing empirical research – with partial exception of the comparative political culture studies on East Central Europe – is the fact that questionnaires are rarely compatible.

2 Calculations are taken from Österreichisches Institut für Wirtschaftsforschung – WIFO Monatsberichte 1993/5, for OECD-countries, and from World Bank, Alton Project, International Comparison/ICP, adjusted by the authors to more recent figures for East Central Europe

3 To cover their cost of living, people therefore have to resort to other economic activities ranging from multiple remunerated employment, or (occasional) employment in Western Europe, to household production, black-market activities or illegal employment (Rose and Haerpfer 1992 and 1993; Haerpfer 1993). As table 1 shows, all these activities help the majority of households to survive; it is, however, also evident that about a third (H, PL, SK) or a quarter (CZ) of households are unable to cover their cost of living.

4 See Fessel+GfK, Geldwesen und Institutionenvertrauen in Ost-Mitteleuropa (1993). The Czech data differ somewhat from those of other countries. It should be noted, however, that the Czech Republic is the only country where unemployment is low even by Western standards.

5 In Russia, 43% of respondents approve privatization in general, while 35% disapprove of it and 21% are indifferent (GfK-Moscow, Social-Political Survey, n = 2.073 respondents in Russia in January/February 1994). For a more detailed discussion of the situation in East Central Europe see Plasser and Ulram (1993b).

6 See Plasser (1992), Plasser and Ulram (1992b) and Fessel+GfK, Lebensstile in Ost-Mitteleuropa (1991 and 1992) and GfK, Haushalts- und Elektro-Panel in Ost-Mitteleuropa (1992/1993).

7 In Austria and Sweden, the importance of cooperativist policy formulation has diminished in recent years (Tálos 1993). In a long term perspective, even in Austria Sozialpartnerschaft seems set to become a (political-) cultural orientation rather than a characteristic element of economic and social policy formulation.

8 Operationalization of "diffuse support" has not yet been solved satisfactorily. Diachronic trend analysis, in addition, requires a replication of internationally standardized indicators for reasons of comparability, which reduces the scope for methodological innovation. The prospects for experimental validation and flexibility in designing questionnaires are further limited by the current shortage of research funds.

9 Contrary to political folklore, which has found its way even into the writings of such distinguished scholars as Samuel P. Huntington (1991, 17), mass democratization in Bohemia and Moravia (the Czech Republic of today) did not begin with the dissolution of the Habsburg Empire but had already got under way some decades earlier. The last decades of the defunct Austro-Hungarian monarchy were characterized by a continuing struggle for constitutional reform and broadening political participation. In the Austrian or Cisleithanian part of the Empire, to which Bohemia and Moravia belonged, universal (male) suffrage was introduced in 1907.

10 Arguably, spoil systems and political interventionism are not unknown in traditional Western democracies, and especially in continental Europe political parties routinely try to exert influence on the media. That kind of influence, however, is generally wielded in a more sophisticated, often institutionalized and thus less arbitrary way. It in no way resembles the experience of a not very distant past, when ruling parties controlled all official channels of information.

11 During Austria's first years of post-authoritarian democracy, a broad consensus of the governing "Grand Coalition" on fundamental economic and political issues was not only important for successful economic performance, but also for the process of political legitimation. The new ruling elites concentrated on consensus building and political integration of all relevant social groups, confining radical opposition to the fringes of the political spectrum. For a more detailed discussion of the resulting interplay of political and structural innovations, and of the development of Austrian political culture, see Ulram (1990).

References

Ágh, A. (1993), 'The Comparative Revolution and the Transition in Central and Southern Europe', *Journal of Theoretical Politics*, no. 2.

Ágh, A. (1994), 'The Social and Political Actors of Democratic Transition', in Ágh, A. (ed.), *The Emergence of East Central European Parliaments: The First Steps*, Budapest.

Ágh, A. (1995), 'The Paradoxes of Transition: The External and Internal Overload of the Transition Process', in Cox, T. and Furlong, A. (eds.), *Hungary: The Politics of Transition*, London.

Alison, L. (1994), 'On the Gap between Theories of Democracy and Theories of Democratization', *Democratization*, no. 1.

Almond, G. A. and Verba, S. (1989), *The Civic Culture. Political Attitudes and Democracy in Five Nations*, Newbury Park, et al.

Barnes, S. H. (1994), 'Politics and Culture', in Weil, F. D. (ed.), *Political Culture and Political Structure: Theoretical and Empirical Studies*. Research on Democracy and Society, vol. 2, Greenwich.

Beetham, D. (ed.) (1994), *Defining and Measuring Democracy*, London.

Beyme, K. (1994), *Systemwechsel in Osteuropa*, Frankfurt.

Brokl, L. and Mansfeldova, Z. (1992), Von der "unpolitischen" zur "professionellen" Politik. Aspekte der politischen Kultur der CSFR in der Periode des Systemwechsels', in Gerlich, P., Plasser, F. and Ulram, P. (Hg.), *Regimewechsel. Demokratisierung und politische Kultur in Ost-Mitteleuropa*, Vienna.

Burnell, P. (1994), 'Democratization and Economic Change Worldwide – Can Societies Cope?', *Democratization*, no. 1.

Chorrin, Ch. (1993), 'People and Politics', in White, St., Batt, J. and Lewis, P.G. (eds.), *Developments in East European Politics*, Durham.

Crawford, B. and Lijphart, A. (1995), 'Explaining Political and Economic Change in Post-Communist Europe', in *Comparative Political Studies*, no. 2.

Dahl, R. A. (1995), 'The Newer Democracies: From the Time of Triumph to the Time of Troubles', in Nelson, D.N. (ed.), *After Authoritarism. Democracy or Disorder?* Westport.

Diamond, L., Linz, J. L. and Lipset, S. M. (eds.) (1990), *Politics in Developing Countries. Comparing Experiences with Democracy*, Boulder and London.

Diamond, L. and Plattner, M. F. (eds.) (1993), *The Global Resurgence of Democracy*, Baltimore and London.

Diamond, L. (ed.) (1994), *Political Culture and Democracy in Developing Countries*, Boulder.

Di Palma, G. (1990), *To Craft Democracy. An Essay on Democratic Transition*, Berkeley and Oxford.

Easton, D. (1979), *A Systems Analysis of Political Life*, Chicago and London.

Gerlich, P., Plasser, F. and Ulram, P. (Hg.) (1992), *'Regimewechsel. Demokratisierung und politische Kultur in Ost-Mitteleuropa'*, Vienna.

Glaeßner, G.-J. (Hg.) (1994), *Demokratie nach dem Ende des Kommunismus. Regimewechsel, Transition und Demokratisierung im Postkommunismus*, Opladen.

Gluchowski, P. and Zelle, C. (1993), 'Ostdeutschland auf dem Weg in das bundesrepublikanische System', in Plasser, F. and Ulram, P. (Hg.), *Transformation oder Stagnation? Aktuelle politische Trends in Osteuropa*, Vienna.

Hadenius, A. (1992), *Democracy and Development*, Cambridge.

Haerpfer, Ch. (1993), 'Demokratie und Marktwirtschaft in Osteuropa im Trend', in Plasser, F. and Ulram, P. (Hg.), *Transformation oder Stagnation? Aktuelle politische Trends in Osteuropa*, Vienna.

Haggard, S. And Kaufmann, R. P. (1994), 'The Challenges of Consolidation', *Journal of Democracy*, no. 4.

Highley, J. and Gunther, R. (eds.) (1992), *Elites and Democratic Consolidation in Latin America and Southern Europe*, Cambridge and New York.

Huntington, S. P. (1991), *The Third Wave. Democratization in the Late Twentieth Century*, London.

Ilonszki, G. and Kurtán S. (1992), 'Traurige Revolution - freudlose Demokratie. Aspekte der ungarischen politischen Kultur in der Periode des Systemwechsels', in Gerlich, P., Plasser, F., and Ulram, P. (Hg.), *Regimewechsel. Demokratisierung und politische Kultur in Ost-Mitteleuropa*, Vienna.

Ilonszki, G. and Kurtán, S. (1993), 'Schöne neue Welt? Politische Tendenzen in Ungarn 1990-1993', in Plasser, F. and Ulram, P. (Hg.), *Transformation oder Stagnation? Aktuelle politische Trends in Osteuropa*, Vienna.

Kaase, M. (1994a), 'Political Culture and Political Consolidation', in Blommestein, H. J. and Steunenberg, B. (eds.), *Governments and Markets. Establishing a Democratic Constitutional Order and a Market Economy in Former Socialist Countries*, Dordrecht, Boston and London.

Kaase, M. (1994b), 'Political Culture and Political Consolidation in Central and Eastern Europe', in Weil, F. D. (ed.): *Political Culture and Political Structure: Theoretical and Empirical Studies. Research on Democracy and Society*, vol. 2, Greenwich.

Klingemann, H.-D., Mochmann, E. and Newton, K. (eds.) (1994), *Political Research in Eastern Europe*, Bonn and Berlin.

Klingemann, H.-D. (1994), 'Die Entstehung wettbewerbsorientierter Parteiensysteme in Osteuropa', in Zapf, W. and Dierkes, M. (Hg.), *Institutionenvergleich und Institutionendynamik*, Berlin.

Lewis, P. G. (1992), *Democrcy and Civil Society in Eastern Europe*, London.

Mainwaring, S., O'Donnell, G. and Valenzuela, S. J. (eds.) (1992), *Issues in Democratic Consolidation: The New South American Democracies in Comparative Perspective*, Notre Dame.

McIntosh, M. E., and MacIver, M. A. (1992), 'Coping with Freedom and Uncertainty: Public Opinion in Hungary, Poland and Czechoslovakia, 1989-1992', *International Journal of Public Opinion*, vol. 4, no. 4.

McLean, I. (1994), 'Democratization and Economic Liberation: Which is the Chicken and What is the Egg?', *Democratization*, no. 1.

Merkel, W. (1994), 'Systemwechsel: Probleme der demokratischen Konsolidierung in Ostmitteleuropa', *Aus Politik und Zeitgeschichte*, nos. 18-19.

Merkel, W. (Hg.) (1994), *Systemwechsel 1: Theorien, Ansätze und Konzeptionen*, Opladen.

Mestrovic, S. G. et al. (1993), *The Road from Paradise. Prospects for Democracy in Eastern Europe*, Lexington.

Meyer, G. (Hg.) (1993), *Die politischen Kulturen Ostmitteleuropas im Umbruch*, Tübingen.

Miller, A. M., Reisinger, W. M. and Hesli, V. L. (eds.) (1993), *Public Opinion and Regime Change. The New Politics of Post-Soviet Societies*, Boulder.

Montero, J. R. and Torcal, M. (1990), 'Voters and Citizens in a New Democracy: Some Trend Data on Political Attitudes in Spain', *International Journal of Public Opinion Research*, no. 2.

Müller, K. (1995), 'Vom Post-Kommunismus zur Postmodernität? Zur Erklärung sozialen Wandels in Osteuropa', *Kölner Zeitschrift für Soziologie und Sozialpsychologie*, no. 1.

Nelson, D. N. (ed.) (1995), *After Authoritarianism. Democracy or Disorder?*, Westpoint.

Nelson, D. N. (1995), 'The Rise of Public Legitimation in the Soviet Union and Eastern Europe', in ibid (ed.), *After Authoritarianism. Democracy or Disorder?*, Westport.

Neuhold, H., Havlik, P. and Suppan, A. (eds.) (1995), *Political and Economic Transformation in East Central Europe*, Boulder.

Offe, C. (1994), *Der Tunnel am Ende des Lichts. Erkundungen der politischen Transformation im neuen Osten*, Frankfurt and New York.

Olson, D. M. (1993), 'Political Parties and Party Systems in Regime Transformation: Inner Transition in the New Democracies of Central Europe', *The American Review of Politics* (Winter).

Parry, G. and Moran, M. (eds.) (1994), *Democracy and Democratization*, London and New York.

Plasser, F. and Ulram, P.A. (1992), 'Zwischen Desillusionierung und Konsolidierung. Demokratie- und Politikverständnis in Ungarn, der ČSFR und Polen', in Gerlich, P., Plasser, F. and Ulram, P.A. (Hg.), *Regimewechsel. Demokratisierung und politische Kultur in Ost-Mitteleuropa*, Vienna.

Plasser, F. and Ulram, P. (1992b), 'Economic Expectations and Political Confidence: Recent Empirical Findings about the Transformation Process in East-Central Europe, in ESOMAR (ed.), *Seminar on Marketing Integration of East and West Europe: Transition and Evolution*, Esomar publication, Amsterdam.

Plasser, F. and Ulram, P. A.(ed.) (1993a), *Staatsbürger oder Untertanen? Politische Kultur Deutschlands, Österreichs und der Schweiz im Vergleich*, Frankfurt et al.

Plasser, F. and Ulram, P. A. (ed.) (1993c), *Transformation oder Stagnation? Aktuelle politische Trends in Osteuropa*, Vienna.

Plasser, F. and Ulram, P. A. (1993c), 'Zum Stand der Demokratisierung in Ost-Mitteleuropa', in Plasser, F. and Ulram, P. A. (Hg.), *Transformation oder Stagnation? Aktuelle politische Trends in Osteuropa*, Vienna.

Plasser, F. and Ulram, P. A. (1993d), 'Of Time and Democratic Stabilization,' Paper for the WAPOR seminar on Public Opinion and Public Opinion Research in Eastern Europe, Tallina (June 11–12th, 1993).

Plasser, F. and Ulram, P. A. (1994a), 'Politische Systemunterstützung in den OZE-Staaten', *Österreichische Zeitschrift für Politikwissenschaft*, no. 4.

Plasser, F. and Ulram, P. A. (1994), 'Monitoring Democratic Consolidation: Political Trust and System Support in East Central Europe', paper for the XVIth World Congress of the International Political Science Association, Berlin (August 21-25).

Plasser, G. (1992), 'Social Change and Perceptions of Consumers', in ESOMAR (ed.), *Seminar on Marketing Intergration of East and West Europe: Transition and Evolution*, Esomar Publications, Amsterdam.

Pradetto, A. (1994), Die Rekonstruktion Ostmitteleuropas. Politik, Wirtschaft und Gesellschaft im Umbruch, Opladen.

Rose, R. and Haerpfer, Ch. (1992), 'New Democracies Between State and Market'. A Baseline Report of Public Opinion, Studies in Public Policy 204, University of Strathclyde, Glasgow.

Rose, R. and Haerpfer, Ch. (1993), 'Adapting to Transformation in Eastern Europe', Studies in Public Policy 212, University of Strathclyde, Glasgow.

Sandschneider, E. (1995), *Stabilität und Transformation politischer Systeme. Stand und Perspektiven politikwissenschaftlicher Transformationsforschung*, Opladen.

Schmitter, P.C. and Terry, K. (1992), 'The Types of Democracy Emerging in Southern and Eastern Europe and South and Central America', in Volten, P. M. E. (ed.), *Bound to Change. Consolidating Democracy in East Central Europe*, Boulder.

Schmitter, P.C. (1994), 'Dangers and Dilemmas of Democracy', *Journal of Democracy* no. 2.

Sorensen, G. (1993), *Democracy and Democratization*, Boulder.

Szabo, M. (1994), 'Nation-State, Nationalism, and the Prospects for Democratization in East Central Europe',*Communist and Post-Communist Studies*, no. 2.

Tetzlaff, R. (1993), 'Demokratie und Entwicklung als universell gültige Normen? Chancen und Risiken der Demokratisierung in der außereuropäischen Welt nach dem Ende des Ost-West-Konfliktes', in Böhret, C. and Wewer, G. (Hrsg.), *Regieren im 21. Jahrhundert*, Opladen.

Ulram, P. A. (1990), *Hegemonie und Erosion. Politische Kultur und politischer Wandel in Österreich*, Wien.

Volten, P. M. (ed.) (1992), *Bound to Change: Consolidating Democracy in East Central Europe*, New York and Prague.

Waller, M. (1994), 'Political Actors and Political Roles in East Central Europe', in *The Journal of Communist Studies and Transition Politics*, no. 1.

Waller, M. and Myant, M. (eds.) (1994), *Parties, Trade Unions and Society in East-Central Europe*, London.

Weidenfeld, W. (Hg.) (1993), *Demokratie und Marktwirtschaft in Osteuropa*, Gütersloh.

Weil, F. B. (1985), 'A Second Chance for Liberal Democracy', Paper Presented at the Annual Meeting of the American Political Science Association, New Orleans.

Weil, F. D. (ed.) (1993), 'Democratization in Eastern and Western Europe'. *Research on Democracy and Society,* vol I, Greenwich.

Weil, F. D. (ed.) (1994), 'Political Culture and Political Structure: Theoretical and Empirical Studies'. *Research on Democracy and Society* vol. II, Greenwich.

Westle, B. (1994), 'Demokratie und Sozialismus. Politische Ordnungsvorstellungen im vereinten Deutschland zwischen Ideologie, Protest und Nostalgie', in *Kölner Zeitschrift für Soziologie und Sozialpsychologie,* no. 4.

White, St. (1994), 'Postcommunist Politics: Towards Democratic Pluralism?', in *The Journal of Communist Studies and Transition Politics,* no. 1.

2 Lifestyle Concepts and their Application in Comparative Research on Central European Political Culture

Gregor Matjan

Introduction

Applying a lifestyle concept to political culture, I ask why political parties and political elites are currently losing so much credibility and trust among the population, not only in old and established Western democracies but also in post-socialist societies of Central and Eastern Europe. In short, I want to scrutinize what – as a syndrome of several indicators – is commonly referred to as political dissatisfaction or – somewhat misleadingly – "antipolitics" (Konrad 1985, Schedler 1994). New social movements and populist actors – on the left as well as on the right – have been attracting a large potential of partisanship and are able to provide alternative symbolic frameworks for identification. Throughout the history of ideas, several notions of "the political" have competed for defining power in a permanent struggle for the inclusion of certain issues, groups etc. into, or their exclusion from, the political arena (Bourdieu 1985, 23–30). Political culture research also encounters a multiplicity of critical, apathetic or even anomic attitudes which all suggest the conclusion that democracies in general have to fight with a growing lack of support. This has been evident from a broad range of different indicators of democratic "quality" during the last ten years. Until now, general and plausible reasons for this development have not been found. The often invoked "normalization" thesis, according to which democracies will – after all – adapt to American standards, cannot be applied to the post-socialist countries whose people desperately fought for democracy some years ago. Astonishingly, the democratization process in these countries has not been accompanied by adequate levels of satisfaction with emerging democratic institutions and civil societies.

But the lack of explanations for this parallel process of degradation in East and West can perhaps be remedied by a broader understanding of political culture than is commonly found in comparative research. Such a concept needs to connect attitudinal data from mass surveys with such elements as have been analyzed mostly under the heading of "civil society". With the notion of lifestyles, a key for connecting these two concepts is presented here. This should result in an integrative approach which (1) provides a theoretical background of both cultural and political theory, (2) includes not only descriptive or empirical facts but also normative and epistemological considerations, and (3) not only deals with attitudinal data but also with symbols, rituals, socio-political practices and social relations.

The complex result of such an enrichment of political culture by other approaches and theories can be simplified if we take into account that there are only a few basic "ways of life", beyond which more detailed symbolic and ideological constructs and interpretations are possible. Furthermore, these ways of life are not equally distributed in society but tend to establish dominant or hegemonic cultural regimes. This brings to mind the central dilemma of where political culture ends and "dominant ideology" begins. As Archie Brown (1979, 12) puts it, "the validity of the concept of political culture cannot be said to have been fully tested until it has been used in a comparative study of communist states, for if the political cultures of societies which have become communist can be readily moulded into a new shape with old values cast aside, the explanatory value of political culture may reasonably be regarded as marginal."

It is indeed questionable whether the breakdown of socialist regimes has been brought about by a renaissance of persistent "old values". More plausibly, these were preserved only to a very small degree and vanished with socialist attempts to deconstruct all forms of "bourgeois" values and social arrangements, resulting in a fatalist culture embedded in a moderate but comfortable privacy (cf. Smolar 1994, 156). Neither were socialist regimes able to create a "socialist man", nor did "old values" simply persist beneath the superficial impact of propaganda. Rather, the regimes promoted a kind of "homo sovieticus" (Alexander Zinoviev), who was "formed under the conditions of dictatorship, a stunted man who shows no initiative and who submits to conditions created and demands made by the authorities" (Smolar 1994, 156). This does not prove that the explanatory power of political culture concepts is marginal. However, the methodological understanding underlying empirical research on political culture has indeed been proved static, reductionist, and imbued with individualist as well as functionalist biases.

Thus, it becomes clear why political culture research has contributed so little to "transitology" and why it was soon supplanted by revived "civil society" concepts. The difficulties of measuring such dynamic social impacts as individualization are rooted in a methodological individualism blind to the *structural aggregation* of individual processes. The whole of a specific political culture always amounts to more than the sum of its components, i.e., of attitudes of individual members of a society. It is – to speak in terms of theories of dynamic systems – an "emergent phenomenon". Measuring individual attitudes towards abstract objects like "the political system" or "the role of citizens" can definitely not suffice as a description of political culture and its impact on social and political change. Therefore, we need to go beyond Almond and Verba's reductionist approach and turn to older notions of culture as offered by cultural anthropology. I shall quote only one out of thousands of definitions of culture to show how much explanatory power has been lost since then. This "tight and unusually thoughtful" definition (Kroeber and Kluckhohn 1963, 120) originates from Redfield (quoted after Ogburn and Nimkoff 1940, 25): "[Culture is] an organization of conventional understandings manifest in act and artifact, which, persisting through tradition, characterizes a human group." Today, the formula "persisting through tradition" needs to be replaced as a result of the impact of technology which has, to a large extent, substituted traditional agencies by dynamic processes of symbolic exchange (by "communication" in a broad sense). A preliminary definition of culture could now run as follows: Culture is an organization of conventional understandings manifest in act and artifact, which, persisting through dynamic processes of symbolic interaction (communication), characterizes a human group.

36

From ways of life to lifestyles

The term "lifestyle" is not yet part of the established vocabulary of political culture research. It is mainly associated with profane consumer culture, with a superficial styling of products, a "designer capitalism" steadily creating and shaping commodities, communication and even persons, and defining what is currently *en vogue*. "Lifestyle" also connotes glamorous or even luxurious settings and is often associated exclusively with upper/middle class contexts (e.g. "The Lifestyles of the Rich and Famous"). Another line of interpretation focuses on the ephemeral styles of constantly changing youth subcultures.[1]

From a more discriminating perspective, lifestyle represents a category applied by cultural sociologists analyzing the results of individualization processes (cf. Featherstone 1987). As such, the term is commonly viewed by critics as affirmative, politically irrelevant or too fuzzy for a "scientific" approach to cultural phenomena. Even more frequently, lifestyles are seen to be typical expressions of late or postmodern capitalism, of its highly diversified economy and of manipulated markets fragmented by target group strategies in advertisement (cf. Weiss 1989). Therefore, one could easily conclude that it is hard to take advocacy of lifestyle concepts seriously, especially with respect to post-socialist countries where the impacts of individualization and Western consumer culture have only been felt for a few years.

In a different context, however, lifestyles can certainly prove relevant for political culture research. The term itself combines elements of a "way of life" with elements of style, i.e. with symbols and their attached meanings. Ways of life (as "cultures") can be deconstructed into "cultural biases" and "social relations". Following *Cultural Theory* by Thompson, Ellis, and Wildavsky (1990, 2), we can formulate a *compatibility condition* according to which "the viability of a way of life (...) depends upon mutually supportive relationships between a particular cultural bias and a particular pattern of social relations. These biases and relations cannot be mixed and matched at random." As a result, only five basic ways of life turn out to be viable. These are derived from two anthropological dimensions, "grid" and "group". "*Group* refers to the extent to which an individual is incorporated into bounded units. (...) *Grid* denotes the degree to which an individual's life is circumscribed by externally imposed prescriptions" (Thompson et al. 1990, 5).

The five ways of life or "cultures" resulting from these assumptions are then described as hierarchy, individualism, egalitarianism, fatalism, and as that of the hermit, which latter, however, has no relevance here. Unfortunately, these labels provided in *Cultural Theory* are somewhat insufficient. Sometimes they refer to modes of social organization, sometimes to dominant attitudes. To accomplish this set of labels, and following the distinction between cultural bias and social relation, we need to add some missing descriptors: hierarchy (as a way of seeing the world) can be complemented with the notions of traditional community or (institutional) bureaucracy. Individualism can be related to markets or similar networks based on exchange processes among persons. Egalitarianism is strongly rooted in affinities with closely knit groups, "sects", or social movements, and is thus frequently referred to as "expressive individualism" (e.g. by Bellah et al. 1985). Finally, fatalism evokes an image of atomized "mass society" as an aggregation of single individuals who – contrary to individualists – are subject to domination and standardization by other ways of life.

These basic elements of political cultures constitute a plurality always to be found in reality (requisite variety condition). It may happen that certain ways of life or alliances of ways of life become dominant or hegemonic at certain moments in history and/or within a specific context (be it a nation state, an organization, or some other subgroup of a population). In today's Western societies, we are confronted with a symbiotic alliance of hierarchy and individualism, of bureaucracies and markets that form the "system" (Habermas 1984), "establishment" (Thompson et al. 1990, 88), or "center" (Douglas and Wildavsky 1982, 83–101). Hierarchy and individualism form a functional and mutually supportive relationship because, in a liberal democracy, hierarchies need individualism for necessary innovations within their stratified and ritualized organization, whereas individualists need hierarchies to provide them with stable, legally structured settings in order to minimize insecurities. Contrary to Habermas's duality of system and lifeworld, it seems to me more useful to apply the terms "center" and "border", since Habermas excludes all aspects of privacy that go beyond the functional imperatives of economic and political institutions. There is more to market individualism and bureaucracy than its formal, "systemic" properties. They also include a diversity of lifestyle elements and informal events "behind the scene", for instance when politicians, journalists, managers, acclaimed artists etc. meet more or less privately. The center is a cultural, social and political sphere where professional relations and private lifeworlds are closely intertwined.

Under socialist regimes, we were facing a different constellation based on the alliance of individualism and fatalism, and described by Wildavsky (1984, 24) as "authoritarianism". Individualism here stands for "enfranchized nomenclatura" networks that determined the sphere of political power and economic planning. Today they continue their individualist way of life as "radicals" (Smolar 1994, 157–161) who gain most of their profits from privatization (see also Heinrich 1995, 28). Complementarily, the amount of social control, together with Marxist-Leninist propaganda and its mass happenings, formed "fatalist" masses. In early modernity, subjection under technological progress forced people into alienated labour and thus into new and highly commodified dependency relations. In the course of history, such atomized individuals were organized both as a standardized mass of consumers and as uniform members of a working class, and were instilled with either the apathetic attitude of fascism that persons must submit to the will of *Volksgemeinschaft*, or the fatalist belief of Marxism that trust in the dialectics of history will lead them toward a brighter future.

The transition from socialist to democratic regimes at first brought forth a change in institutional structure, but also – more fundamentally and earlier – a process of individualization. This was mainly a result of first cautious attempts at liberalization regarding private economic activities in the late 1970s. Reception of Western television and radio broadcasts was made possible by the ensuing modest affluence and further stimulated the need for individualization and mobility. Arguably, individualization not only triggered the "velvet revolutions" but subsequently also released people from fatalist "mass" arrangements into an individualized existence. Alexis de Tocqueville (1946, II, 100) already noted with respect to the American Revolution that "Individualism [is] stronger at the close of a democratic revolution than at other periods". If we apply a lifestyle model of political culture, a certain progress of individualization would even seem to be one of its main conditions, since it requires that people be in a position to practice symbolic distinction and choose their identities. This, however, would be incompatible with the conformity of a highly standardized mass society. With respect to politics, contingency of adherence entails an opportunity to decide not only between a growing number of political parties (pluralization), but also to choose certain ways of life which include a particular understanding of what politics generally is, how it should work, and who should participate in which agenda by what means. Democratization must allow for very subjective and often ambivalent practices of shaping individual political affiliations and activities that need not be homogenous, or even consistent by their internal logic or ideology. Individualization released people from "embedded" arrangements of penetrating party structures, political "camps" and milieus. Today they tend to refuse to think in functional categories with respect to the political system since they no longer see themselves as part of it. The unfolding of individualism must therefore be seen as a major force in the shaping and transforming of a society's institutions, organizations and political culture.

Lifestyles: a matter of rational choice?

Every day, we are confronted with a multiplicity of ways in which people try to organize and sort out their lives. It has emerged as a central feature of so-called postmodern societies that there are virtually no restrictions on the appropriation of different aspects of culture to assemble an individual lifestyle, on making life a *bricolage* of social, political, religious, cultural and other elements (Hitzler 1994). On the basis of relatively secure affluence and social welfare, we are allowed to make choices. Traditional communities allegedly no longer restrict the way we behave, dress, think or vote. We choose our party identification in the same way we choose our brand of toothpaste at the supermarket – in a highly commodified context, it makes no difference whether we select and buy a product or vote (cf. Soeffner 1989). For conservative critics like Daniel Bell (1976) and the late Aaron Wildavsky (1991), this freedom of choice ends up in a "psychedelic bazaar" where nobody takes responsibility for anybody else, where work ethics and values of achievement become undermined by expressive individualism, egalitarianism, political correctness, and affirmative action programmes. In their eyes, such "cultural contradictions of capitalism" lead to economic, political, and moral decay.

Nevertheless, societies do not exactly follow this pessimist vision of arbitrariness. New social arrangements emerge from the ruins of old communities. Forced to maintain a high degree of mobility, people reorganize their affiliations by seeking already existing models of social arrangements into which they can fit. Undergirded by communal development strategies, this leads to cultural "cluster" building on regional and urban levels (gentrification, gerrymandering). To refer to this process as individualization, however, is misleading, since the latter does not consist in a linear, constant and non-reversible move away from mechanic *Gemeinschaft* to organic *Gesellschaft*. Rather, individualization is a dynamic and even chaotic restructuring of societies. In constructing new life contexts, we create epistemological frames that help us reduce the complexity of our surroundings. These cannot be compared to social systems in the understanding of system theory, with autopoietic information circles and strict boundaries. Lifestyles, as I call such frames, are temporary, ephemeral, and fuzzy. What makes them so important for culture in general, and for political culture in particular, is that they are not simply given, but "enacted" through our practices of perception (Varela

1992). In this constructivist view, we do not simply perceive the world by creating representations of sense impressions inside our brain, but conceive it through our problematization of certain aspects of the reality we have to cope with. We develop routines, focus upon selected contexts, and narrow our perspective to concentrate on a problem that needs to be solved. In an interpretation of broader scope, this process of enacting does not happen individually, but as a "social construction of reality" (Berger and Luckmann 1966). Through communication, we learn to relate our individual epistemological frameworks to those of a "generalized other" (Mead 1964, 284). At higher aggregate levels, certain processes become ritualized and based on manifest social structures; "institutions" emerge. As such, they constitute functional elements for integration and reproduction of society.

In Western literature on the subject, it is sometimes claimed that the freedom to choose a way of life rationally is a fundamental precondition for an application of lifestyle models (e.g. Lüdtke 1989). This appears to be of special relevance for an adaptation of the concept to Central Europe: during the period of transition, most private and public social security systems and networks broke down, not least due to individualization. Social inequalities between the new rich and masses subsisting beneath the poverty line has increased rapidly within the last years. It might seem highly cynical to talk of freedom of choice, when, for a large number of people, sheer survival is their first and foremost goal. However, choosing a particular lifestyle may be a possibility even under materially distressing living conditions. For instance, there are marked differences in the way people arrange their lives in cases of homelessness. Even in this situation, cultural practices of distinction are part of someones management of poverty (cf. Thompson and Wildavsky 1986). When viewed from an "objective" point of view, there is almost no opportunity for substantial choice wherever somebody gets trapped in a downward spiral of poverty. But it could also be argued that even people with "objective" opportunities of choice at their hands may feel severely restricted. Since the perception of possible alternatives is subjected to collective cultural restrictions (some things one simply "cannot do"), choosing a lifestyle is a very complex matter that goes far beyond the limits of rational decisions.[2] Theories insisting on the rationality of choice thus exclude problems of preference formation which can only be grasped epistemologically or normatively.[3] The imputation of "rational" decisions remains quite simply meaningless if it is not related to cultural constraints.

A normative framework for political culture

The concept presented here tries to connect the notions of political culture and lifestyle. It allows for a synopsis of peoples' subjective constructions of meaning in accordance with material, cultural, and educational constraints that constitute the "objective" framework for subjective attitudes and practices. A lifestyle concept should enable political culture research to formulate an integrative approach which is less eager to discover new analytical distinctions and catchwords (such as "postmaterialism" or *Politikverdrossenheit*), but rather tries to isolate particular collective syndromes of rituals, symbols, attitudes and social relations combined in a way of life. From a book on interior design, I take the following "unscientific" description of lifestyle, which nevertheless demonstrates the heuristic and integrative quality of this term:

"A number of factors – economic, cultural, material, and societal – influence the spirit of a lifestyle. But in the long run, perceptions and attitudes change; reactions occur towards the taste of a bygone era" (Chic Simple 1993, 15).

By emphasizing the search for subjective identities, lifestyles convey a strong normative claim relevant for political culture, a claim rooted in the idea of authenticity as moral imperative: everybody should live his or her life according to his or her ideas of the "good life". Charles Taylor (1991) has criticized individualist premises of authenticity norms allegedly indicative of a trivialized culture. I do not share this communitarian critique since, as Taylor himself concedes, there can be no notion of authenticity derived simply from the individual self. He seems to be blurring the distinction between liberal (utilitarian) and egalitarian (expressive) individualism. The former is neither able nor willing to formulate collective visions, while the latter conceives of authenticity as the shared practice of distinction. Concepts of the good life, visions, and utopias someone wants to live up to are societally formed and communicated in various discursive contexts. Also, it is impossible to live a good life in solitude (except, to some extent, for hermits), and without communicative exchange or ideas and actions that affect others. Authenticity needs always to be proved to "generalized others" though it appears primarily to be a claim individuals address to themselves. There are, however, some fundamental assumptions connected with authenticity which enable us to pronounce moral judgments concerning an individual's conduct of life.

One such principle has been called "weak relativism" by Taylor. It asserts that all attitudes and ways of life should have the right to exist without interference from others. Due to individualization, cleavages cutting across societies have multiplied, but they no longer reach as deep as they used to. The claim to authenticity (comparable to the one layed down in the American Constitution, and according to which every man should be entitled to pursuit of his own happiness) becomes a general feature of postmodern societies. Weak relativism may not provide highly individualized societies with a strong organizing principle, but it does allow for a certain openness with respect to cross-cultural communication and mutual understanding. A multicultural society consisting of more than ethically defined groups primarily requires a loose and sometimes quite chaotic organizing principle that avoids forcing people into closed structures of singular identities. Another, much stronger principle derives from the need for recognition (Taylor 1992). Contrary to the individualist bias of weak relativism, a claim to recognition can only be formulated by a community, by a group of people sharing a common identity, but not simply by adherents of a way of life. The implications of this communitarian vision turn out to be highly problematic due to the mainly metaphysical line of argument that ignores practical intricacies, such as social closure of identity-based communities.

In the model presented here, "politics of self-actualization" or "life politics", as Giddens (1991) has called this phenomenon, claims the realization of collective identities as personal lifestyles. With respect to political representation, these claims reach far beyond the classic institutions of representative democracy. Nowadays, the quality of democratic representation is not so much assessed by formal criteria such as exact proportional parliamentary shares, as by the degree to which representatives of "the people" (politicians) behave authentically. That is, they will be judged morally by their ability or inability to accord public statements, visions, or policy measures with their private lifestyle. Consequently, public and private spheres conflate, since performance in one sphere will be measured by the representatives' behavior in the other. By way of example, one could mention members of the "Tuscany faction" within the German Social Democrats (SPD) who lost credibility in the eyes of the public by displaying a hedonistic lifestyle which did not match that of many SPD voters. One can find similar discrepancies in every democratic political system, be they hard corruption, soft lies, phony propaganda or simple ignorance.

From the perspective of a lifestyle model of political culture, political dissatisfaction is rooted in the artificial division politicians erect between the realms of the private and the public, particularly if behavior in the two spheres turns out to be incompatible and does not meet the public demand for authenticity. A crisis of representation and declining confidence in political elites logically result from the chasm of lifestyles between an authenticity-seeking public and the symbolic politics provided by its representatives (Edelman 1964). Phony rhetoric or symbolic politics that represent little else than propaganda will soon be unveiled by mass media and lead to distrust, public contempt, civil disobedience, or anomic reactions by large sectors of the population.

While symbolic politics have consistently become discredited, political symbolism on the other hand is a typical feature of lifestyle-politics from "below". Lifestyles can be seen as latent cultural patterns situated within the frame of the four basic ways of life provided by cultural theory. In addition to the anthropological dimensions of "grid" and "group", another attribute is of importance here:[4] the human urge to make sense of surrounding by means of symbols. Symbols are used for distinction, differentiation, and to attract attention. For political culture, the two following modes of symbolization appear to be of special relevance.

Sapir (1930, 492ff) has described such symbols as "referential" and "condensational". Examples for referential symbols are national ones such as flags, anthems, and state emblems, but also party logos, slogans or even individuals who help us identify certain abstract entities (cf. Edelman 1964). The particular quality of these symbols lies in their general communicability. They can draw on a common understanding, a generalized meaning, a shared interpretation, and are open to critical testing against reality.

Condensational symbols, in contrast, do not have a general meaning. They are encoded in relation to the receivers' competence in decoding (cf. Hall 1993). Apart from their restricted communicability, they are also more emotionally laden than referential symbols. Sapir has specified this kind as "condensational symbols" because they are often used to focus disparate meanings and messages for the purpose of gaining legitimacy, partisanship, or identification.

In Western countries, new social movements – green-alternative as well as New Right – have had a strong impact on political cultures by focusing on condensational symbols. During the transition from socialist to democratic regimes, condensational symbols played an important part in boosting the self-confidence of civil rights movements. The "Solidarnosc" logo, to be sure, elicited a degree of recognition, but it very soon – and primarily – came to symbolize the unity of the movement and attracted support and solidarity from all over the world. Whether symbols are instrumentalized from "above" or from "below", by political elites or by parts of a civil society, depends mainly on the purposes they serve.

Disseminated by mass media, symbolic politics are a modern instrument of propaganda conveying the idea of a demarcation between public and private spheres, while political symbolism tries to do the reverse. By symbolizing personal concern, social movements strive for a practice of authenticity which carries supposedly private

42

matters of personal lifestyles into the sphere of "the political". The political realm is thereby extended, and the established institutions and their elites are challenged. We should note, however, that it is not only (mostly egalitarian) activists who resort to this kind of symbolization, but also passive, apathetic "consumers" of politics. As a result of their disappointment with politics, the latter practice "conspicuous consumption" and act on their preference for material goods or TV sitcoms that closely symbolize their frustration with political matters. Following Watzlawicks "metacommunicative axiom" (1967, 1), according to which it is impossible not to communicate, we can maintain that within a lifestyle society it is impossible not to symbolize anything.

Towards a political theory of culture

The conventional approach to political culture shows little interest in explaining cultural phenomena. It rests on the old definition given by Almond and Verba according to which political culture consists in the "particular patterns of orientations *toward* political objects" (Almond/Verba 1963, 14f). This interpretation of political culture was borrowed from the democratic theory of David Easton and his systemic perspective on politics. Within such a functional systems approach, political culture appears only as a residual category within a political system that consists of institutions, parties, and other functional elements (cf. Gerlich 1984, 21). Political culture is reinforced by the output of the political process, but has no autonomous qualities. From Parsons (et al., 1953) onwards, systems theory conceptualized culture as a functional element of societies, as "latent tension management and pattern maintenance". This notion of culture contains a fiercely reductionist and functionalist bias. Because it dismisses the notion of culture as a structure underlying all other functional relations in society, political culture research has lost much of its explanatory capacity, at least as far as the impact of socio-structural changes like individualization on the democratic process are concerned.

> Only when we have an understanding of the making of culture, and of the organization of access to it, can we talk coherently about the interpretation put upon and the actions derived from it. The immediate problem is, therefore, not just one of developing a cultural theory of political interests, but a political theory of culture (Street 1993, 113).

Since Almond's and Verba's seminal comparative study on *The Civic Culture*, the notion of methodological individualism has helped to establish the assumption that the democratic "quality" of a political culture can best be measured by following rules similar to those governing the democratic process: one man – one vote. This initiated a paradigmatic shift in political culture research toward a fully "empirical" science, a development accomplished by Ronald Inglehart's longitudinal studies on value change in Western societies. But when societies enter rapid transformation processes – as it happened in socialist East Central Europe – empirical methods reach the limits of their applicability. It turns out to be almost impossible to measure individualization using a methodology like empirical survey research based on "methodological individualism". Conventional political culture approach has so far rarely been able to predict the outcomes of such processes, be they spontaneous students riots in 1968 or "velvet revolutions" in 1989. Obviously, other factors beside the "particular patterns of orientations *toward* political objects" must have played a part in these dramatic changes within political cultures. Would it not, perhaps, be more fruitful to view political culture as the cultures *of* political actors, *of* institutions, or – more generally – *of* collective arrangements?

Introducing lifestyles as catalytic converters between "objective" structural conditions and "subjective" ways of coping with them, we can see political culture as a set of common practices of problematizing and symbolizing certain parts of reality considered politically relevant. What exactly is meant by "political" must be left open; if we take authenticity to be a serious but trivialized claim (Taylor 1991) implying an ineliminable plurality of interpretations concerning the nature of this "political", then neither political science nor philosophy can define it universally and in advance. A phenomenological or inductive approach to political culture will be necessary if we are to grasp the various interpretations and understandings of politics. We need to be aware of the fact that different traditions of political thought have established a durable plurality of such understandings. At various points in history one paradigm came to dominate or eclipse another, but alternative interpretations were always at hand. Most recently, the debate between communitarians and contractarian liberals provides a perfect example of this plurality of political language games. Within political cultures, redefinitions of the political are happening all the time, although in a trivialized and somehow arbitrary form. But the task of political culture research should be to relate such popular struggles for defining power to elaborate political theories and ideas.

The four ways of life outlined in *Cultural Theory* might give us a rudimentary impression of what "constrained relativism" (Thompson et al. 1990, 269) with respect to definitions of the political could look like:

Individualist notions of politics have been formulated by prominent theorists like Schumpeter, Downs, Popper, and most classical liberal authors. (Democratic) politics appear as highly rationalized, market type procedures, in which everyone is busy maximizing his or her utility. Political elites know best how to manage a state, and should therefore be not interfered with by too far-reaching participatory claims. Thus, Mancur Olson and other game theorists have analyzed the conditions under which collective action comes about.

Hierarchical conceptions of society arose with the bureaucratization of the early modern state. German *Verwaltungswissenschaft* (science of public administration) was created and perfected during the 19th century and served as a model for Max Webers ideal type of rational government founded on law. Even today, the most widespread perception of politics is that of a hierarchical order of aggregated interests on the one hand, and of executive bureaucratic hierarchies on the other. From this perspective, politics looks like a bundle of impersonal institutions ordering society. Even modern systems theory in the versions of Luhmann, Willke and others, while (counterfactually?) denying hierarchy, retains hierarchical notions of politics in its insistence on functional differentiation.

In contrast to these approaches, *egalitarian* notions of the political offer a critical perspective on existing institutions and governing elites. Prominent theorists of egalitarian politics are Rousseau, Tocqueville, Arendt, and communitarian thinkers like Charles Taylor or Michael Walzer. In their critique of hierarchies and markets, egalitarians support the inclusion of as many groups as possible (universal suffrage) and strongly advocate an active civil society and social movements. At the center of their political designs, however, a highly moralized vision of the political with correspondingly exclusionary tendencies is to be found.

Finally, *fatalist* visions of the political have rarely been formulated in modern times, although we can find some prominent exponents of such notions in Thomas Hobbes and the utilitarian thinkers. According to Hobbes, man is unable to control his passions and only a strong "Leviathan" will guarantee peace and order. Marxism and theories of "mass society" like Adorno's and Horkheimer's "Dialectics of Enlightenment" (1947) analyzed the subjection of individuals under dominant ideological powers incisively; both, however, relied on the fatalism of ineluctably dialectic historical processes determining the political struggle of mankind.

These language games of the political show a plurality exceeding the spectrum of programmes offered by political parties. Competing democratic parties are necessarily attracted by the "center" of different cleavages cutting across the public space, and the struggle for a majority of votes leads to a homogenization of their agendas. In a political culture structured by lifestyles, however, imageries of the political are linked to imageries of the social. How individuals perceive politics depends on their way of seeing the world at large. Politics in this sense is nothing other than a strategy of problematizing different aspects of the world. Where no problems occur, no politics is needed. There are, however, different strategies of coping with problems: ignorance, indifference, diffuse antipathy, taking risks, etc. Due to the fragmentation of societies, social imageries and symbols have become highly diversified. The emergence of new political parties and civic organizations on the left and right (and also liberal) ends of political spectra is a good indicator of this diversification.

A lifestyle approach to political culture

If we want to go beyond an interest in political culture as merely a variable explaining the legitimacy of political systems, we need to ask how and why specific patterns of political culture occur, and how they can be explained by structural variables. In research on social inequalities, lifestyle approaches have recently gained in importance, mainly because the disparity between the realms of the subjective and the objective has proved an intractable problem. Due to individualization, older models of class or social stratification have lost more and more of their prognostic capacity regarding the influence of social structure on (political) culture or any pattern of behaviour (cf. Street 1993). In the search for "key structures" that link material conditions and cultural expressions, lifestyles are increasingly coming to be recognized as such by students of political culture. As key functions, they allow us to loosely connect and bracket people with different material backgrounds on the basis of common symbolic codes. Moreover, they provide epistemological frames of reference by which others are judged.

Lifestyles also serve as means of distinction and identification. As such, they help us reduce the complexity of the world around us, which is, of course, generally the main function of "culture". From a phenomenological perspective, we could even say that they are symbolically integrated and subjectively created lifeworlds. In their strong insistence on authenticity, they also exhibit a moral dimension: we should be honest to ourselves and not confront others with expectations we could never live up to ourselves. In this way, public and private spaces conflate since everyone's public statements and actions may be judged by his or her private lifestyle.

The special relevance of lifestyles for political culture is rooted in the structural homology of parties and lifestyles. From the point of view of political sociology, we could say that parties (in a broad sense) are institutional structures which both provide political identification for their members and participate in a constant struggle for symbolic distinction from their competitors in the political field of parliamentary democracy.

> Every political discourse is in some sense split from within. While apparently aimed at the clientage concerned, the *de facto* addressees are competitors in the field (Bourdieu 1985, 35).

As in my preliminary definition above, lifestyles may be seen as self-organized pre-political entities based on symbolic exchange. People intending to associate will first need to effectively judge each other by their appearance and practices of symbolic distinction. Before they enter into communication, mutual expectations need to be at least partially fulfilled. It thus becomes necessary to use correct symbolic codes to be in turn assessed correctly by others, to avoid, in other words, permanent misunderstanding. Much of a person's behavior can be symbolically interpreted, regardless of the fact that this distinctive behavior need not be pursued consciously.

> Distinction does not, as is often assumed following Veblen's theory of *conspicuous consumption*, imply a conscious attempt at distinction. Any act of consumption or, more generally, any practice is *conspicuous*, is visible, irrespective of whether or not it was performed with a view to its *being seen* (Bourdieu 1985, 21).

Practices of symbolic distinction not only reveal an individual's position in a stratified society, but also his or her level of education, identification with a certain age group or generation, cultural preferences and, to some extent, political creeds and affiliations (Bourdieu 1984). As Douglas and Wildavsky (1982) have shown, a central issue for political culture research with respect to distinctive practices are strategies of dealing with risks. This has been confirmed by Ulrich Beck in his best seller *Risikogesellschaft* (1986), where he relates the social dynamics of individualization to the cultural dynamics of risk perception. Beginning in the 1980s, ecological issues were at the center of the debate on risk and risk avoidance. Nowadays, the risk agenda has widened to include all aspects of complex social processes. Even individualization itself is problematized as risky.[5] At the same time, counter-strategies of risk avoidance are evolving, the most important being new nationalist and regionalist movements (from Canada to Scotland, from Italian Leagues to Slovak and Romanian Hungarians, etc.).

Is such a lifestyle approach to political culture restricted to the analysis of Western democracies? If not, to what extent can it be applied to Central and Eastern Europe? My argument is based on the fact that the transition to democracy forced most people living in these countries to reposition themselves within the new social and political order. But when individuals are released from settled life contexts, exposed to standardization by Western consumer culture, and atomized by individualist principles of political and social organization, they face a paradoxical situation of having to choose. The same claim to the freedom of choice, which was one of the strongest motives impelling the revolutions behind the Iron Curtain, is now turning into a problem for people trying to reconstruct their social lives. In their attempt to redefine aspects like privacy, family, home, national affiliation, political orientation, sources of information, and so on, they cannot easily resort to old values, institutions, and circumstances of before 1938 or 1949. Too much has been lost to continue bourgeois or even peasant forms of life, even though the rebirth of some such social and political traditions (like peasant parties) turned out to be viable, if nostalgic, options. A sense of uncertainty and confusion seems now to dominate the restructuring of social relations and cultural moulds. No one quite knows how to integrate the new, mainly Western influences interfering with old habits of a relatively sheltered lifestyle promoted under the former regime. Under socialism, managing uncertainties was never taught, and today the pressure for adaption is quite simply too much for many people.

We know from theories of nationalism that nations can be seen as "imagined communities" (Anderson 1983). But the need to imagine or (re)invent a community always arises in the absence of an existing one, i.e. under conditions of atomization or individualization. This is why nationalism appears to be intertwined so closely with modernity (Gellner 1983). Individualism as a way of life requires not only a highly developed competence in networking (one can never rely upon communities) but also a willingness to take high risks (cf. Douglas and Wildavsky 1982). But under uncertain legal conditions, for example, taking risks might seem so hazardous as to commend a strategy of symbolic integration in order to reduce the complexity of one's surrounding.

Strategies of risk aversion can take several forms: in "expressive" individualism, people participate in egalitarian citizens' action groups trying to reduce risks (social and ecological "postmaterialism"); in mass societies, risks are dealt with by reassurance through symbolic identification (fatalism). Glorification of history, nationalist rhetoric, and a stereotypical construction of enemy camps characterize the fatalists' search for security and strength at a time when such conditions do not in fact prevail. While strategies of risk aversion have ceased to be peripheral phenomena in Western societies and are being applied persuasively and massively by neopopulist actors, such groups have been much more clamorous in East Central Europe. They have also revealed the scale and significance of this syndrome. Nationalist and national-communist movements rely heavily on symbolism, on images of a glorious past (be it communist or pre-communist) and on icons representing these bygone times. Disseminated by the media, condensational symbols appeal to much larger audiences than could ever be gathered in public spaces (e.g. at public rallies), since they are presented, with propagandist artifice, as though referring to some ascertained and unquestionable truth.

These strategies fall on fertile ground wherever social and cultural circumstances – the predominant lifestyles – make people responsive to offers of this kind. Those who are used to making choices in daily life and take more or less severe risks all the time, are less inclined to adopt crude symbolic dichotomies of good/bad, inside/outside, etc. In "common sense" language, we could say that someones attitude toward the process of individualization will largely depend on his or her "horizon", although we would do well to remember that epistemic formulae like "horizon" are closely tied to, and reinforced by, social relations. In the face of pluralization and social fragmentation, nationalist movements offer larger symbolic orders as reactions against (or even antidotes to) dynamics of individualization which are changing fatalist lifestyles more and more rapidly. Thus, it becomes almost impossible to interpret a political culture on the basis of lifestyles without using a dynamic approach. One such promising methodological and theoretical framework is offered by what is commonly referred to as "chaos theory". Chaos theory enables us to conceptualize the emergence of macro-stability through micro-variability not in a simplistic and functional manner, but as a

47

continual reciprocity of different ways of life, cultures, and lifestyles (cf. Bühl 1987, 71 and Jancar 1992).

Conclusion

Can, therefore, a lifestyle approach be usefully applied to the analysis of Eastern European political culture? And if so, what would its advantages be compared to empirical surveys of political culture? We have seen that the dominant mode of social development in the reform countries can be described as one of individualization. Parallel to this dynamics on the micro level, new democratic conditions allow for further pluralization resulting in the emergence of a highly differentiated civil society. With the legal, technological, and material basis for wider access to diverse media in place, a highly complex landscape of symbolic codes will soon be a reality. If we renounce on the prerequisite of personal freedom of choice for reasons elaborated above, the conditions for a shaping of lifestyles would be acknowledged and a justification of lifestyle analysis provided. Nevertheless, I would like to caution against applying this concept to East Central European countries on the following grounds:

It is apparent that the speed and direction which change in these societies takes is specific to each country, and varies greatly with different traditions and infrastructures. Consequently, common features dwindle in importance. At the same time, the imperatives of democratization, privatization, and modernization are overwhelming and will become ever more dominant. But we cannot predict today how social relations will develop under these circumstances. Some countries (the Czech Republic, Hungary, and Slovenia) will soon be indistinguishable from Western countries as far as the extent and depth of individualization and the emergence of a lifestyle political culture is concerned, while in others (Croatia, Slovakia, and Poland) this process will be strongly countervailed by traditional, communitarian, nationalist, and religiously motivated politics. Analyses based on individualization and lifestyles in the latter case would be almost absurd, and would certainly require a huge constructivist effort. In these countries, neopopulist political actors form a strong alliance with those "integrative" parts of society that represent a nationalist, risk avoiding, and fatalist culture trying to withstand the allurements of modernity, individualization, and liberality. As the past has proved, individualization has outlasted socialism. The future will show whether these regressive political cultures can prevail under dramatically altered circumstances.

Notes

1 See the research carried out by the proponents of "Cultural Studies", e.g. Hebdidge (1979).
2 As understood by subjective utility theory: concrete strategic alternatives with expected payoffs or outcomes exist, and the demand for sufficient information is met (cf. Riker and Ordeshook 1973, 21–23).
3 See the critique by Wildavsky (1994).
4 Reference to anthropological properties of humanity does not imply strong assumptions of a "state of nature", or similar discursive figures. In this context, "anthropological dimensions" denote relatively unproblematic traits frequently observed in the structuring of societies.
5 Beck has analyzed this discourse as "reflexive modernization".

References

Almond, G. A. and Verba, S. (1963), *The Civic Culture Political Attitudes and Democracy in Five Nations*, Princeton.

Anderson, B. (1983), *Imagined Communities: Reflections on the Origin and Spread of Nationalism*, London.

Beck, U. (1986), *Risikogesellschaft. Auf dem Weg in eine andere Moderne*, Frankfurt.

Bell, D. (1976), *The Cultural Contradictions of Capitalism*, New York.

Bellah, R. N. et al. (1985), *Habits of the Heart. Individualism and Commitment in American Life*, Berkeley.

Berger, P. L. and Luckmann, T. (1966), *The Social Construction of Reality*, Garden City, N.Y.

Bourdieu, P. (1984a), 'Espace social et génèse de classe', *Actes de la recherche en sciences sociales*, no. 52/53, June 1984.

Bourdieu, P. (1984b), *Distinction. A Social Critique of the Judgment of Taste*, Cambridge.

Brown, A. (1979), 'Introduction', in: Brown, A. and Gray, J. (1979), *Political Culture and Political Change in Communist States*, New York, pp. 1–24.

Bühl, W. L. (1987), *Kulturwandel. Für eine dynamische Kultursoziologie*, Darmstadt.

Chic Simple (1993), Home (ed. by Kim Johnson Gross and Jeff Stone), Munich.

Douglas, M. and Wildavsky, A. (1982), *Risk and Culture. An Essay on the Selection of Technical and Environmental Dangers*, Berkeley et al.

Edelman, M. (1964), *The Symbolic Uses of Politics*, Urbana, Ill.

Featherstone, M. (1987), 'Lifestyle and Consumer Culture', *Theory, Culture & Society*, no. 4, pp. 55–70.

Gellner, E. (1983), *Nations and Nationalism*, Ithaca, N.Y.

Gerlich, P. (1984), 'Parlamentarische Kultur', *Österreichische Zeitschrift für Politikwissenschaft*, no. 1, pp. 21–26.

Giddens, A. (1991), *Modernity and Self–Identity*, Cambridge.

Habermas, J. (1984), *Theory of Communicative Action*, 2 vols., Boston.

Hall, S. (1993), 'Encoding, decoding'; in During, S. (ed.), *The Cultural Studies Reader*, London, pp. 90–103.

Hebdidge, D. (1979), *Subculture – The Meaning of Style*, London.

Heinrich, H.-G. (1995), 'Postsozialistische Individualisierungsprozesse. Probleme der 'Zivilisierung' osteuropäischer Gesellschaften', *Kurswechsel*, no. 1, pp. 21–35.

Hitzler, R. (1994), 'Reflexive Individualisierung. Zur Stilisierung und Politisierung des Lebens', in Richter, R. (ed.), *Sinnbasteln. Beiträge zu einer Soziologie der Lebensstile*, Vienna et al., pp. 36–47.

Horkheimer, M. and Adorno, T. W. (1947), *Dialektik der Aufklärung. Philosophische Fragmente*, Amsterdam.

Jancar, B. (1992), 'Chaos as an Explanation of the Role of Environmental Groups in East European Politics', *Green Politics Two*, Edinburgh, pp. 156–184.

Konrád, G. (1985), *Antipolitik. Mitteleuropäische Meditationen*, Frankfurt.

Kroeber, A. L. and Kluckhohn, C. (1963), *Culture. A Critical Review of Concepts and Definitions*, New York.

Lüdtke, H. (1989), *Expressive Ungleichheit. Zur Soziologie der Lebensstile*, Opladen.

Mead, G. H. (1964), *Selected Writings*, ed. A. J. Reck, Indianapolis.

Ogburn, W. F. and Nimkoff, M. F. (1940), *Sociology*, Boston.

Parsons, T. et al. (1953), *Working Papers in the Theory of Action,* New York.

Riker, W. H. and Ordeshook, P. C. (1973), *An Introduction into Positive Political Theory,* Englewood Cliffs/N.J.

Sapir, E. (1930), *'Symbolism', Encyclopedia of the Social Sciences,* New York, pp. 492–495.

Schedler, A. (1994), 'Antipolitical Opposition. A Framework for Comparative Analysis', unpublished paper presented at the Vienna Dialogue on Democracy, Vienna.

Smolar, A. (1994), 'Die samtene Konterrevolution', *Transit 8,* pp. 149–170.

Soeffner, H.-G. (1989), 'Die Inszenierung von Gesellschaft – Wählen als Freizeitgestaltung', in Haller, M., Hoffmann-Nowottny, H.-J. and Zapf, W. (ed.), *Kultur und Gesellschaft,* Verhandlungen des 24. Deutschen Soziologentags, des 11. Österreichischen Soziologentags und des 8. Kongresses der Schweizerischen Gesellschaft für Soziologie in Zürich 1988; Frankfurt and New York, pp. 329–345.

Street, J. (1993), 'Political Culture – from Civic Culture to Mass Culture', *British Journal of Political Science,* no. 24, pp. 95–114.

Taylor, C. (1991), *The Malaise of Modernity,* Concord, Ontario.

Taylor, C. (1992), *Multiculturalism and the Politics of Recognition,* Princeton.

Thompson, M. and Wildavsky, A. (1986), 'A Poverty of Distinction: From Economic Homogeneity to Cultural Heterogeinity in the Classification of Poor People', *Policy Sciences,* no. 19, pp. 163–199.

Thompson, M., Ellis, R. and Wildavsky, A. (1990), *Cultural Theory,* Boulder, et al.

Tocqueville, A. de (1946), *Democracy in America,* 2 vols., New York.

Varela, F. J. (1988), 'Whence Perceptual Meaning. A Cartography of Current Ideas' , in Varela, F.J. and Dupuy, J. (ed.), *Understanding Origins. Contemporary Views on the Origin of Life, Mind and Society,* Dordrecht et al., pp. 235–236.

Watzlawick, P. (1967), *The Pragmatics of Human Communication,* New York.

Weiss, M. J. (1989), *The Clustering of America,* New York.

Wildavsky, A. (1984), *The Nursing Father. Moses as a Political Leader,* Birmingham, Ala.

Wildavsky, A. (1991), *The Rise of Radical Egalitarianism,* Washington, D.C.

Wildavsky, A. (1994), 'Why Self-Interest Means Less Outside of a Social Context: Cultural Contributions to a Theory of Rational Choices', *Journal of Theoretical Politics,* no. 2, pp. 131–159.

3 The Symbolic Dimension. Political Anthropology and the Analysis of ECE Political Cultures

Andreas Pribersky

In their contributions to this volume, a number of authors question the possibility, purpose, and methodological premises of measuring political culture. They raise, in particular, the issue of whether an international, comparative perspective can be applied to a group of states which have only recently appeared on the political map of Europe and emerged from the first stages of a rapid change in their legal, institutional, and political frameworks. Yet another difficulty in evaluating East Central European political cultures arises from the fact that the concepts used in survey questionnaires and comparative studies are interpreted differently in each of the countries concerned. Thus it could be argued that the scope for misinterpretation widens with a given political culture's divergence from the standard comparative model. In this way, historical and/or cultural differences between East Central Europe and Western (European) countries may be mirrored in divergent interpretations of particular indicators of political culture, and in the significance these are attributed within a comparative context.

A well-known example of such interpretive differences between quantitative similarities in comparisons of political cultures is the decline in political participation: a trend recognized in Western European democracies for at least a decade, and whose main indicator has been taken to be the continuously increasing abstention from voting. Most of the new ECE democracies began to exhibit a similar trend soon after the first waves of democratization, i.e. after the first free elections had been held. There is, however, a growing consensus among analysts that an apparently comparable degree of voting abstention may have very different reasons in, respectively, Western and East Central European democracies, and that an explanation based primarily on comparative data may be seriously misleading (see also Attila Ághs essay in this volume).[1]

Moreover, public debates following the publication of comparative political culture surveys in ECE media suggest that their accuracy or correspondence with self-images prevailing among citizens of the country may not be as close as imagined. Studies of this kind derive the remarkable attention paid to them from the fact that, along with economic growth data or inflation rates, indicators of political culture (such as turnout at elections) are seen as defining a country's relative position in ECE competition for integration into Western (European) political and economic structures. Especially in signatory states of the Visegrad agreement (Hungary, Poland, Czech Republic and Slovakia), the public awaits quantitative comparisons with the expectation or hope that they present "one's own" country more favourably than others. Debates in the media and elsewhere are accordingly limited to deliberations on the implications and consequences of non-negotiable norms of democratic political culture. When Hungarians, according to a recent survey, proved more pessimistic about their economic future than citizens of any other ECE state, Budapest dailies argued that this could discourage potential foreign investors unaware of the fact that pessimism is characteristic of Hungarian culture and need not affect economic behaviour. Although this example is, of course, as incidental in scope as it is anecdotal in intent, it may nevertheless serve as an illustration of cultural differences in the approach to comparative concepts.

Most references to political culture in ECE – even by political scientists – betray a strong urge for, and belief in, democratic "normalization" as a final stage of transition. Not only is such a view of the transformation process part of the official stance in matters concerning EU integration; it also indicates the near-unanimous public validation of a universalism constitutive of comparative studies. The objective of a "normal" democracy implies the assumption of socialization processes involving both the political elite and the population, and converging towards universally shared democratic norms and values[2]. To finalize transition to democracy in this way, however, almost inevitably means that differences between particular democratic cultures will be minimized and brushed aside as short-lived symptoms of transition. But the implicit understanding that there is, if not one high road to democracy, then at least one *telos* of Westernization and democratic reform, cannot even do justice to the variability of Western political cultures. By no means does it take a descriptive approach to recognize this heterogeneity (cf. Plasser and Ulram 1993), since European democracy as a label applies to such diverse political systems as Switzerland's consociational government and France's (or Poland's) presidential system.

A descriptive approach

Descriptive approaches to political cultures are the result of growing criticism concerning the attempt to lay down a universal model of democracy and derive from it the parameters for quantitative comparison and measurement of democratization processes. Beginning in the 1970s, definitions and concepts used in political culture research have been reconsidered from a variety of – psychological, Marxist or systems-theoretic – perspectives (Gibbins 1990, 4f).

Two recurrent features of these debates over methodology have proved particularly relevant with respect to East Central Europe: the possibility or impossibility of universally valid definitions of democratic political culture, and the charge of ethnocentrism levelled against such definitions (Gibbins 1990, 6ff). Defining comparative values of political culture might well constitute a methodological exclusion or reduction of differences in interpretation (as shown in my examples above) and thus eliminate specific cultural patterns from the analysis of political cultures; and since applying a putative golden mean of Western political experience to the new European democracies might lead us to overlook their particular problems and dynamics, the charge of ethnocentrism seems reasonable enough.

As a result, research on political culture undertaken within the methodological framework of political anthropology aims primarily at relativizing ethnocentric views of democracy. In their reappraisal of Almond and Verba's concept of "civic culture"[3] as a prerequisite for democratic stability, Thompson, Ellis and Wildavsky show that, in traditional democracies as well as post-war "newcomers", such stability can be achieved through a variety of different, historically and culturally rooted values and attitudes. According to their typology of political mentalities, democratic systems may either rely on a predominantly "individualistic orientation on the group-level" and competitive lifestyle, or on a more egalitarian set of values; most likely, however, they will exhibit a unique combination of the two (Thompson et al. 1990, 247ff). This anthropological typology of political cultures results from a holistic description of a given culture's main features and social patterns. Not only does it allow for a comparative analysis of the implications such patterns might have for political attitudes; it also opens up a perspective on the relativity and particularity of norms and values governing democratic political cultures. In so doing, it provides an excellent starting point for the analysis of these attitudes.

Instead of an approach based on political opinions expressly stated in party programs, political statements, or opinion polls and aimed at the comparative analysis of, for instance, explicit messages or legal and institutional structures, political anthropology focuses on implicit or explicit social rules governing, and evident in, the construction of such messages and institutions. Whereas quantitative research on political culture interprets opinions on certain values, political systems or issues, political anthropology tries to either reconstruct dominant (or redundant) group identities and the political attitudes they imply (Wildavsky 1994, 150ff), or to describe the rituals and symbols used by such groups in the construction of their identity. The latter approach turns to processes of political identification occurring during election campaigns, to political ceremonies, national commemorations, and to the auto-(re)presentation of governments or parties in order to reconstruct national patterns of political culture (Augé 1994, 81ff and Mach 1993).

The daily routine of political (re)presentation expressed in these rituals must be recognized as a main concern for contemporary politics and politicians, especially as presentation in mass media has long become crucial in our so-called electronic age (Balandier 1992, Edelman 1964). Increasing "personalization of politics" and attention to images should also make the analysis of their construction an important task for political culture research (Pekonen 1990, 138f). This is perhaps even more true for societies engaged in a transformation of their political systems. Symbolic politics, or the use of symbolic communication for the purpose of linking the identification processes of social groups with politicians aspiring to "represent" them, allow us to analyze dominant trends within a political culture. In the following examples focusing on paternalism as a characteristic feature of ECE political cultures, I will try to outline the scope and possibilities of an analysis of political culture in its "symbolic dimension".

Paternalist patterns

Paternalism has often been recognized as a major problem of Central European political cultures, and its prevalence has been confirmed by comparative data analyses covering the region.[4] In its origin, the importance attributed to paternalism for an explanation of ECE political attitudes coincides with the attempt, increasingly noted during the 1980s, to (re)define a "third region" located in the interstices between East and West.

This debate was in the main led by dissidents (see Ash 1986) and tried to attach a political significance to differences in the social structures of the Soviet Union – where civil society is said to have no historical roots whatsoever – and some of its ECE satellites (Czechoslovakia, Poland and Hungary) where Soviet rule was forced to accept compromises between single-party rule and the differentiated societies in place (Vajda 1989).

To substantiate this distinction between three European regions – an image of Europe which, in many ways, influences current geopolitical reasoning and EU integration policies – the Hungarian historian Jenő Szűcs (1990) has tried to reconstruct historical patterns of power-sharing between political elites and societies. He sees the West as having formed societies at a very early stage, and the East as never having struggled free from an all-dominant state; the shaping of Central Europe has been a combination of the two, with the state playing a more dominant role than in the West, but allowing for the development of an increasingly independent society. Szűcs sees this region as approximately defined by the borders of the Habsburg empire, and as marked by political elites instrumental in the creation of independent societies. This setup, he argues, has deeply ingrained the paternalist pattern through centuries of modernization from above.

Paternalism has become a key concept not only in historical approaches to social structure, but also in attempts to evaluate recent social and political developments. Mihaly Vajda (1989) describes the starting point of democratic transformation in Hungary, Poland and the former Czechoslovakia from the perspective of changes in, and of, political elites (which seem to have returned after the last parliamentary elections in Poland or Hungary). He thereby offers a perspective on transition differing from the image of a popular mass movement implicit, for instance, in Ash's term of "refolution", and mirrors the traditional leading role of political elites in the region.

Following Vajdas view of transformation, the process of democratization in the region could similarly be described as a paternalistic attempt at modernization and Europeanization from above. If we turn, as Pekonen (1990) suggests, to the symbols used in the continuous construction of elite images, paternalism may be assumed to emerge as a central pattern in the web of symbolic languages that make up political identification processes in East Central Europe.

In their analysis of symbolic language employed during Hungary's parliamentary election campaign of 1990, Agnes and Gábor Kápitány (1991) show that paternalistic symbols used in the presentation of the Hungarian Democratic Forum (HDF) and its leader Joszef Antall may well have contributed substantially to their success. HDF's main slogan "A nyugot érö" presented both as "The Stable Force" in the process of transformation. In conjunction with Antall's image of a stately *pater patriae*, this responded to many voters's quest for security in times of reorientation and rapid change. Antalls image – in contrast to that of his party – even remained more or less untarnished throughout the many political quarrels and until his premature death in office as prime minister.

In this respect, the slogan "Government by Experts" coined by the Hungarian Socialist Party (HSP) during its successful 1994 campaign, as well as its focus on a single leader, shows remarkable similarities to the staging of HDF in 1990 (Kapitány 1995). The image created of the party and incumbent prime minister Gyula Horn during this campaign may perhaps best be described as a modernized version of paternalism. It addressed the electorate's desire for modernization – in contrast to HDF's strategy of championing traditional "national" values as a source of stability – as well as their expectation that such modernization should come from above. HSP thus opted for a strategy of meeting a set of attitudes that, in 1989–90, had already been addressed by the last socialist government of the old regime, and in which Horn served as Minister of Foreign Affairs.

In a way, this Hungarian example of paternalism as a prime moving force in electoral behaviour seems to confirm the results of quantitative research on Central European political cultures. But it also questions the very idea of normalization and universalization of democratic norms, since the differentiation of political camps along ideological lines – as the changeover from an HDF-led conservative coalition government to the current leftist-liberal coalition has commonly been interpreted[5] – seems to be less significant compared to the preference for paternalist leadership expressed through both choices.

The paternalist pattern and its impact on Hungary's democratic elections since 1990 might serve to illustrate the difference in results obtained by quantitative and descriptive analyses of Central European political cultures. Even though there are no detailed analyses for other, similar cases, several characteristic features of ECE politics seem to fit the perceived pattern. Take, for instance, the power struggle between Slovak president Kovacs and prime minister Mečiar, which may be seen as a clash between rivalling claims to true representation, personification, or even embodiment of "national interest".[6] Or consider the recent presidential election campaigns in Poland, during which both Walensa as incumbent and Kwasniewski as main challenger likewise presented themselves as personifying national interests – thereby repeating a historical pattern in Polish politics dominant since the 1980s and in the struggle for legitimacy between a "state party" and its opposition (Mach 1993, 163ff).[7]

While attitudes such as paternalism thus seem to have appreciable reverberations in many fields of politics, their impact on the development of political culture can hardly be quantified since they operate on a level of symbols beyond the immediate reach of, for instance, explicitly stated ideological preferences. The role of mass media in processes of political identification in general, and their ever increasing importance in election campaigns especially, call for a wider perspective on political culture; one, that is, which takes account of symbolic discourses and their use (Pekonen 1990, 134ff). This seems even more important in the case of East Central Europe, where the pace of change has resulted in a chronic shortage of comparative data and made reliable interpretation difficult. As a result, quantitiative interpretation – of, for instance, comparative data on paternalism – must in part take a descriptive approach. But such borrowings will remain superficial unless they are set in the kind of interpretive context provided by political anthropology.

Research on political cultures in ECE: some perspectives

Of course, an analysis of social symbols and their use in politics alone cannot suffice as a description of ECE political cultures. But it might indicate the direction further analysis will need to take if it is to provide satisfactory definitions of concepts used in survey research on political cultures. While it can thus serve to complement quantitative analysis, the description of symbolic politics also offers its own cross-cultural perspective. Mach (1993, 95ff) has compared national identities of political systems as different as those of Britain, Israel, Poland and the former USSR; in so doing, he has also demonstrated ways of making comparative research more attentive to particulars. Further and more differentiated research could, for instance, investigate the self-images and roles of ECE members of parliament by describing everyday institutional routine – as Abélès (1992) has done in his typology of the European Parliament.

In conclusion, to approach political culture through different fields of symbolic politics seems particularly called for with respect to Eastern Europe or any country undergoing rapid changes in its political system. In all such cases, an analysis restricted to quantification of – invariably scarce – data is bound to neglect cultural patterns of *longue durée* as well as any non-explicit reference to culturally dominant political symbols. In contrast, descriptive studies of decision-making or parliamentary routine could allow for insights into the direction of ongoing democratization processes. It could also take account of, and do justice to, the differentiation in national political cultures developing within comparable legal and institutional framework.

Notes

1 Similar considerations apply to the issue of low confidence in political parties.
2 See Ágh's or Mesežnikov's contributions to this volume.
3 Almond and Verba's notion of, and research on, "civic culture" having provided a point of departure for subsequent interest in political culture guided by a universalizing intent (see Gibbins 1990, 6f).
4 See Plasser and Ulram's, Mesežnikov's or Novosel's articles in this volume.
5 For an example of this interpretation see again Ághs contribution.
6 See. again, Mesežnikov's article in the present volume.
7 Helga Hirsch in Die Zeit, 3 November, 1995.

References

Abélès, M. (1992), *La vie quotidienne au Parlement européen*, Hachette, Paris.
Augé, M. (1994), *Pour une anthropologie des mondes contemporains*, Aubier, Paris.
Ash, T. G. (1990), *We the People – the Revolution of 89 Witnessed in Warsaw, Budapest, Berlin and Prague*, Granta Books, London.
Ash, T. G. (1986), 'L'Europe centrale existe-t-elle?', *Lettre International*, no. 10.
Balandier, G. (1992), *Le pouvoir sur scènes*, Balland, Paris.
Edelman, M. (1964), *The Symbolic Uses of Politics*, University of Illinois Press, Urbana.
Gibbins, J. R. (1990), 'Contemporary Political Culture: An Introduction', in Gibbins, J. R., *Contemporary Political Culture*, London, pp. 1–30.
Kápitány, A. and Kápitány, G. (1991), 'Politikai szimbolumok', *Hungarian Political Yearbook*, pp. 235–241.
Kápitány, A. and Kápitány, G. (1995), 'Az 1994 – es választási kampányfilmek szimbolikus és értéküzenetei' *Hungarian Political Yearbook*, pp. 112–126.
Mach, Z. (1993), *Symbols, Conflict and Identity. Essays in Political Anthropology*, New York State University Press, New York.
Pekonen, P. (1990), 'Symbols and Politics as Culture in the Modern Situation: the Problem and Prospects of the "New"', in Gibbins 1990, pp. 127–143.
Plasser, F. and Ulram, P. (eds.) (1993), *Staatsbürger oder Untertanen? Politische Kultur Deutschlands, Österreichs und der Schweiz im Vergleich*, Peter Lang, Frankfurt etc.
Szűcs, J. (1989), *Die drei Regionen Europas*, Frankfurt.
Thompson, M., Ellis, R. and Wildavsky, A. (1990), *Cultural Theory*. Westview Press, Boulder.

Vajda, M. (1989), *Orosz Szocializmus Közép-Europában*, Századveg, Budapest.

Wildavsky, A. (1994), 'Why Self-Interest Means Less Outside of a Social Context: Cultural Contributions to a Theory of Rational Choices', *Journal of Theoretical Politics*, no. 6, pp. 131–159.

SLOVENIA

4 Politics: Power Struggle or Quest for the Common Good? Slovenian Public Opinion on Politics

Adolf Bibič

With the establishment of party pluralism and parliamentary democracy in Slovenia, the question of what Slovenes think about politics has gained new relevance. Especially following the recent political upheavals and the experience of incipient parliamentary democracy in the new context of a sovereign nation state, it is important to find out how politics as a distinctive activity is perceived by those it involves and concerns. Since 1990, several SPO polls[1] have sought to address such concerns by eliciting views of the public on newly emerging political practices and institutions (democratization, multiparty elections, parties, parliament, government, political actors, etc.); others have been aimed at gauging attitudes towards politics more generally. However cautious we may have to be in drawing inferences and in determining implications of surveys for a general picture of Slovenian political behaviour, the data are invaluable for any understanding of the country's political culture.

There are at least two ways of using opinion surveys as sources for political culture research. One is to ask questions about particular aspects of political life – e.g. institutions, actors, activities or issues – in order to reconstruct a set of political attitudes by means of inference from particulars. Such inferences will be the more plausible, the greater the diversity of aspects covered, provided they yield a coherent picture. The second strategy is to address general attitudes directly and as such, rather than via the detour of details. It may be argued, of course, that both approaches should complement each other and are necessary for a fuller understanding of political attitudes.

The Slovenian Public Opinion (SPO) project has mostly adopted the first approach, and the second in some cases. Given SPO's long term scope and the great emphasis it has placed from the very beginning (1968) on numerous, chiefly political issues, abundant material on attitudes towards various aspects of politics has been collected over the past quarter of a century. The structure and dynamics of these attitudes have also been analyzed (see Toš, 1987, 1989 and 1992). As part of this project or in connection with it, several more detailed studies on Slovenian political attitudes have been undertaken (Markič 1991; Tomc 1993).

What, then, do Slovenes think about politics? In the following, I will attempt to analyze the SPO 1993/1 survey, which dealt with general views on politics and tried to elucidate comprehension as well as evaluation of *de facto* politics.

On the basis of concepts taken from political theory and political science, as well as on that of widespread popular notions of politics, we proceeded from the following hypotheses concerning comprehension and evaluation of politics:

1. politics will be seen as a struggle for power;
2. politics will be seen as a quest for the common good;
3. politics will be seen as a battle for supremacy between special interests;
4. In all probability, politics will be valued negatively rather than positively.

The following questionnaire containing four general descriptions of politics and allowing for six levels of agreement/disagreement on each item was designed:

Table 1 Perceptions and judgment of politics in Slovenia (SPO 1993/1)

We will read to you some opinions on politics. Please indicate which one you agree with, and to what extent.

	agree entirely	agree in the main	unde- cided	in the main, do not agree	do not agree	do not know
a) Politics is a struggle for power	44.7	34.0	5.7	4.4	1.2	10.1
b) Politics is a quest for the common good and fundamental values	5.8	22.1	18.4	25.2	12.9	15.5
c) Politics is a battle for supremacy between special interests	29.1	40.0	9.1	5.7	1.2	14.9
d) Politics is a dirty business that honest people avoid	30.2	29.8	15.0	11.7	3.4	10.0

"Politics is a struggle for power"

A vast majority of respondents agreed with the statement that politics is a power struggle. More than three quarters of them (78.7% or 821 of N = 1054) approved entirely or in the main. More respondents agreed decisively than moderately. The ratio of those "agreeing entirely" and those "agreeing in the main" is 4:3. Most Slovenes, then, see politics or political activity as the business of conflict, and the exercise of power as its objective, the term "power" obviously connoting, in this context, the struggle for influence on national policy-making. Slovenes thus derive their understanding of politics largely from empirical political realities of pluralist parliamentary democracy in its public or other manifestation. We should remember, however, that widespread perception of politics as a power struggle is not the same as (moral) approval with this concept of politics.

Interesting results are obtained when this view of politics is correlated with education, occupation, gender and age of respondents. Among the factors positively correlated with the view that politics is a power struggle, mention must first be made of education. Across categories ranging from incomplete primary school to tertiary or higher education, we may say that the higher the level of education, the greater the proportion of respondents who view politics entirely or in the main as power struggle: the share increases from 65.2% in the bracket of "primary education" to 92.8% in that of "tertiary education" (Table 2). With respect to occupation, respondents with tertiary education are again those who most frequently perceive politics as a power struggle (91.3%) [11.11%][2], surpassed only by 100% of tradesmen or entrepreneurs employing others, who account, however, for only 1.9% of the sample. Next on the list are employees with secondary schooling (85.9%) [17.4%], and artisans or entrepreneurs not employing anyone (80.1%) [0.9%]. The smallest share of affirmative answers is found among unskilled workers (66.7%) [6.0%] and farmers (67.8%) [2.1%]. Interestingly, no significant gender differences can be detected in responses to "politics is a struggle for power" (82.7% of men exceeding women by only 7%). In contrast, age reveals a higher degree of correlation: 80% [21.9%] of respondents aged 30 years or younger saw politics as a power struggle, a view found still more frequently (85.7%) [18.4%] among respondents between 31 and 40 years old, but gradually falling to 67.6% [11.2%] among those 61 years or older.

"Politics is a battle for supremacy between special interests"

Slovenes' prevailing view of politics as struggle and conflict is also reflected in responses to the statement that "politics is a battle for supremacy between special interests". Although affirmative responses were fewer (9.6%), even here the degree of consensus is remarkable: just under three quarters (69.1%) of interviewees agreed. Politics thus conceived extends the principle of struggle for influence and domination on a state level to all spheres of society. It is seen as a distributive hassle determining "who gets what, when and how", as a contest for the larger share of material wealth, organizational power, media presence, interest group domination, presence in upper echelons of parties and the bureaucracy, etc. Implied in this concept is a more sociological understanding of politics as a process resulting from the deep structures of society, or as multilayered political occurrences forming something French political scientists like to refer to as *la vie politique*. Of course it would be implausible to suggest that respondents were aware of all these dimensions of politics. However, the fact remains that a great majority did qualify politics as an activity grounded in the diversification of, or competition among, interests and interest groups, rather than simply as one of political parties competing for power.

Table 2 "Politics is a struggle for power" and education

	agree entirely	agree in the main	total	% of total sample
incomplete primary	37.4	27.5	64.9	5.7
completed primary	32.2	33.0	65.2	14.6
Incomplete junior secondary	31.0	45.2	76.2	3.0
Completed junior secondary	44.0	38.7	82.7	22.8
Completed secondary	50.2	34.5	84.7	19.1
Incomplete junior tertiary	58.5	31.7	90.2	3.5
Completed junior tertiary	62.7	28.8	91.5	5.1
Completed tertiary	69.6	23.2	92.8	4.9
No answer	50.0	25.0	75.0	0.3

With only minor deviations, there is a close correlation between education and the view that politics amounts to a battle for supremacy between special interests: the share of those agreeing entirely or in the main rises with the level of education. The highest percentage of affirmative answers within any one educational category is found among respondents with completed secondary (88.2%) [5%] and tertiary education (85.7%) [4.7%]. Correspondingly, the lowest percentage may be found among respondents with incomplete or completed primary schooling (50.6% [4.4%] and 52.8% [11.7%] respectively). Leaving aside tradesmen and entrepreneurs due to their relatively small share in the sample, the highest incidence of affirmative answers is found among employees holding university or other higher education degrees (85.7%) [10.3%] followed by those with secondary education (81.6%) [16.6%] and skilled workers (72.3%) [15%]. It is considerably lower among unskilled (45.3%) [4.1%] and semi-skilled workers (50.9%) [5.3%], farmers (61.3%) [1.9%] ranging somewhere between those two groups. With respect to gender, there are no substantial differences although once again men (71.1%) lead women (66.7%). Age, in contrast, is a major factor in responses to the statement heading this chapter: respondents aged 50 years or younger perceive politics as a battle between special interests to a greater extent (70 to 72%) than those older than 50 years (65 to 63%).

"Politics is a quest for the common good"

So far, these largely concurrent findings would suggest that most Slovenes agree rather more strongly with the assertion that politics is in fact a struggle for power or social and political dominance. It will now be interesting to see how they react to the idea of politics as primarily an activity oriented towards achieving the *bonum commune* and welfare of all. Thus table 1 shows that the proportion of respondents conceiving of politics in this way is markedly lower than that of people inclined to the two views discussed above. The statement that "politics is a quest for the common good and fundamental values" was confirmed entirely or in the main by just over one quarter of respondents (27.9%). In other words, the goal of general interest supposedly fundamental to politics is accepted by only a minority of Slovenes. Moreover, this perception of politics relates asymmetrically to the above alternatives in that the ratio between decisive and moderate agreement with "politics is a quest for common good" is the reverse of that found in affirmative responses to "politics is a power struggle". Only 5.8% are positive that politics is a quest for the common good, the remainder (22.1%) are moderately convinced at best, the ratio between the former and the latter being approximately 2:7. Slovenes' incredulous attitude towards the idea of politics as "the quest for the common good" is also obvious in the large share of negative answers: as many as 38% of respondents entirely or in the main disagreed with the proposition. In comparison, only 5.6% of respondents disagreed entirely or in the main with the idea that "politics is a power struggle" and only 6.9% with the statement that "politics is a battle for supremacy between special interests".

It will first be noted that the differences between respondents concerning this statement are less pronounced, ranging from 22.4% to 33.7%. With respect to education, the largest share of positive responses to the idea of politics as a quest for the common good is found among interviewees with completed junior secondary (33.7%) and incomplete junior tertiary (31.7%) [1.3%] education. It is slightly smaller in the brackets of completed junior tertiary (28.8) [1.6%] and completed tertiary education (26.7%) [1.1%]. The lowest correlation between education and a view of politics as aiming at *koinonía* was found in the category of individuals with completed primary school (22.4%) [5%] and incomplete junior secondary education (23.8%) [1%]. With respect to occupation, the highest rate of agreement with the proposition is found among the self-employed (50%), although again it must be noted that they accounted for only 0.1% of the sample. Next come employees with tertiary education (32.6%) [4%], and people never having held any kind of employment (32.2%) [3.6%], skilled workers (31.5%) [6.5%], artisans or entrepreneurs employing others (30%) [0.6%], employees with secondary education (27.4%) [5.5%] and, finally, farmers with 25.9% [0.8%]. The percentage of agreement regarding politics as a quest for the common good is lowest in the category of semi-skilled (21.3%) and highly-skilled workers (19.2%) [2,2%]. Once again there are no major gender differences, while once again men take the lead (31.9% as against 24.6%). The closest correlation is found with the age of respondents: the older the respondent, the less likely he or she is to agree with the notion of politics as a quest for the common good. The turning point seems to be situated somewhere around the 30th birthday: up to that time, twice as many respondents agree with the proposition than between the age of 31 and 40 years; and they are even three times as likely to agree as people aged 60 years or older (see Table 3).

Table 3 Politics as a quest for the common good and age of respondent

	up to 30 years	31–40	41–50	51–60	over 60
completely agree	18.5	5.4	5.2	3.4	4.6
agree in the main	51.8	24.1	18.5	22.8	17.3
total	70.3	29.5	23.7	26.2	21.9
% of total sample	9.4	6.4	4.8	3.8	3.7

Although the above questions dealing with the general notion of politics were intended to elicit a descriptive, rather than an evaluative, understanding of politics, it is implausible to assume that value judgements did not colour the responses. This is particularly evident with regard to the statement that "politics is a quest for the common good", although it is just as visible in the statement that "politics is a power struggle" or "a battle between special interests". SPO 1993/1, moreover, contained a question designed to elicit value judgments in connection with politics (see Table 1). This was considered expedient since previous surveys and general observation had indicated a growing dissatisfaction with politics. In this respect, too, Slovenia has been catching up with traditional democracies.

"Politics is a dirty business that honest people avoid"

The above statement could be expected to provoke unambiguous responses and provide a basis for judging how Slovenes evaluate politics. Although these evaluations are of course not uniform, they do confirm the hypothesis that politics is predominantly valued negatively. As Table 1 shows, 60% of respondents agree with the view that politics is, indeed, a dirty business, the internal distribution of decisive and moderate agreement being 1:1. Just over 15% of all respondents disagree with the negative judgement. But this percentage is too small to redress the overall balance. Nor do the residual 15% "undecided" or 10% "do not know" responses affect the general picture.

Among the different educational categories, agreement with the view that politics is, as Goethe said, "a dirty song", dominates among respondents with junior secondary (63.1%) [17%], completed secondary (61.7%) [13.9%] and incomplete junior secondary (62%) [2.4%] education. It should be noted that these three categories alone make up a good 35% of the total sample. The lowest incidence of negative judgements of political activity as such occurs in the category of incomplete (63.9%) [1.7%], completed junior tertiary (54.2%) [3%], and tertiary education (50%) [2.6%]. A breakdown according to occupation shows that negative evaluation is expressed by 65% [1.3%] of artisans and entrepreneurs employing others, by 64.8% [6.7%] of semi-skilled workers, 61.5% [3%] of highly-skilled workers, 61.3 [1.8%] of farmers, 55.6% [6.8%] of employees with tertiary education, and 61.3% [12.5%] of employees with secondary education. Once again, no substantial gender-linked differences were found, while women lead men in their negative judgement of politics by a margin of just over 1%. With regard to age, the lowest share of respondents viewing politics as a dirty business is found among respondents aged 30 years or younger (57.7%) [15.8%], the highest in the brackets of "41–50 years" (64.9) [13.1%] and "60 years or older" (63.6%) [10.6%]. The age groups of "31 to 40 years" (56,3%) and "51 to 60 years" (59.1%) fall somewhere in between. People no older than 40 years, therefore, tend to have a better opinion of politics than their seniors.

Conclusions

1. Slovenes have a relatively poor opinion of politics. This emerges clearly from the SPO 1993/1 as well as other opinion surveys.

2. This generally negative view is also confirmed by attitudes to particular aspects of politics. Confidence in political parties is declining rapidly; trust in central political institutions is also eroding, although to a lesser extent; and the performance (efficiency) of administrative authorities is given a low rating (with the exception of the president, the government, and the Ministry of Defence).

3. Therefore, Slovenia is already saddled with what German political scientists and politicians refer to as *Politikverdrossenheit* (disenchantment with politics).

4. The following causes of such negative views may be identified, without pretending to any hierarchichal ordering:

   – the general view of politics as a distinctive social activity which, by definition, endows it with certain 'Machiavellian' features defying ordinary morality;
   – in particular, the fact that politics is in part synonymous with conflict has not become part of political culture;
   – Slovenes have never before had an independent sovereign state, which means politics has largely been identified with foreign rule up to very recently;
   – sensationalism in the media encourages a habit of personalizing and merely touching on issues in a rapid succession, thereby impoverishing the public image of politics;
   – political scandals and their misuse for party infighting have damaged the reputation of politics;
   – the elite's political culture has left a lot to be desired in the initial years of transition. Through the media, these deficiencies begin to loom large in the eyes of the public.

The damaging impact of dissatisfaction with politics has been recognized even for well-established democracies (Beyme 1994). It is likely to be even more harmful in a young country and democracy like Slovenia, which is engaged in a process of constructing a new national and political identity. In this context, excessive disillusionment with politics could easily lead to a type of apathy stalling this process at its outset. Of course, indifference and resignation also leaves a vacuum routinely filled by demagogic populism. As experience in traditional democracies has shown, an aversion to political involvement may open the door to political extremism, which today mainly has a right wing bent (compare Italy, France, Germany, Austria and elsewhere) but would necessarily fuel countervailing left wing extremisms in the long run. Along with internal stability, the strength and standing of the young Slovenian state – perhaps even its territorial integrity – could be imperiled.

A balanced reading of the results of the SPO survey and even the SPO 1993/1 does, however, allow for a somewhat more optimistic interpretation of Slovenes' attitudes to politics. It may be noted that the history of the Slovenian people is also that of a millenium of collective survival amid powerful and not always benign neighbours: a history that has recently resulted in sovereign statehood. All important elements of cultural and political tradition in Slovenia, ranging from Christianity to socialism and liberalism, have contributed something to this development. Despite the predominantly negative designation of politics as "dirty business", Slovenes have begun to accept pluralist, parliamentary democracy along with its defects. This is evident from the relatively high value ascribed to elections as instruments of political democracy and as channels through which political decisions may be influenced. The high turnout at nationwide elections confirms that the public does indeed grasp their significance and assumes a degree of responsibility in building the new state. A state, incidentally, whose independence they support almost without reservation.

Slovenes have, therefore, been developing more differentiated attitudes towards politics. This may be inferred also from the fact that political awareness has extended to a reevaluation of Slovenia's role during World War II. There have, of course, been symptoms of national selfishness and intolerance, though on a relatively moderate scale. But public support for openness towards the world in general and Western Europe in particular by far outweighs countervailing tendencies. Thus it could be said that the spatial as well as the temporal range of Slovenes' political awareness has been growing steadily, together with critical attention directed at obstacles to Slovenia's EU integration. Slovenian public opinion has also taken a turn in the direction of a secular political culture, revealed most clearly in attitudes concerning the Church and its political role.

SPO polls indicate that Slovenes tend to approach politics with pronounced realism – as a power struggle – and that they are critical of it insofar as it takes the form of a fierce struggle resorting to unfair means and thus violating the rules of the game. The relatively high proportion of presondents who qualify politics as dirty business probably stems mostly from a critical stance towards this kind of power struggle. Such a normative approach to politics comes to the fore even more clearly in responses to the statement that politics seeks the common good. Less than a third of respondents opted for this notion, but other responses to more specific aspects of politics suggest that, on a normative level, this view is much more widespread. Further research will probably reveal that the idea of politics as a common interest is also implicit in the assertion that politics is in fact a power struggle or a struggle between special interests. The decided rejection of particularist party activities and the marked espousal of certain values which imply an understanding of politics as a quest for the common good (solidarity, national autonomy, positive evaluations of coalition policies, etc.) already corroborate this assumption. But more continuous, historically aware and comparative research will be needed before such speculations can be validated.

Notes

1 Slovenian Public Opinion Surveys are conducted by the Centre for Public Opinion and Mass Communication Research at the University of Ljubliana's Faculty of Social Sciences.
2 Percentages in square brackets refer to the total sample (N=1054).

References

Beyme, K. (1994),'Politikverdrossenheit und Politikwissenschaft', in Leggewie, C., *Wozu Politikwissenschaft*, Wissenschaftliche Buchgesellschaft, Darmstadt.

Conolly, E., and William, E. (1983), *The Terms of Political Discourse*, Martin Robertson, Oxford.

Krišnik, Z. (1994), *Slovenski politiki izza pomladi narodov* (Slovenian Politicians after the Spring of Nations), Frafa, Ljubljana.

Markič, B. (1992), 'Politika in javno mnenje' (Politics and Public Opinion), *Teorija in praksa*, nos. 1–2.

Pleterski, J. (1994), 'Pravica in moč. Slovenci in država' (Right and Might. Slovenes and the State), *Naži razgledi*, no. 21, pp. 14–16.

Tomc, G. (1993), 'Slovenci o politiki in politikih' (Slovenes on Politics and Politicians), in Adam, F., *Volitve in politika po slovensko* (Elections and Politics in Slovenia), Znanstveno in publicisticno sredisce, Ljubljana.

Toš, N. (ed) (1987*), Slovensko javno mnenje. Pregled in primerjava rezultatov raziskav SJM 68–SJM 1987* (Slovenian Public Opinion. Review and Comparison of Survey Results SPO 1968 to SPO 1987), Delavska enotnost, Ljubljana.

Toš, N. (ed) (1992), *Slovenski izziv. Rezultati raziskav javnega mnenja 1990–1991.* (The Slovenian Challenge. Public Opinion Survey Results 1990–1991), FDV–IDV (Public Opinion and Mass Communications Research Centre), Ljubljana.

Toš N. (ed) (1994), *Slovenski izziv II. Rezultati raziskav javnega mnenja 1992–1993* (Slovenian Challenge II. Public opinion research results 1992–1993), FDV–IDV (Public Opinion and Mass Communication Research Center), Ljubljana.

5 Political Culture in a Context of Democratic Transition. Slovenia in Comparison with other Post-Socialist Countries

Danica Fink-Hafner

My purpose here is to discuss some theoretical approaches to political attitudes and political participation in post-socialist countries. I will also present empirical data on political values and behaviour in East Central Europe in general, and in Slovenia in particular; special emphasis will be placed on the relationship between parties (political elites) and voters (society), as well as on indicators of anti-party sentiment and other sets of values and attitudes relevant for political participation.

The politics of post-socialism: some concepts

Theoretical approaches to democratic transition and its actors are offered by modernization theories, by research into Southern European democratization since the 1970s, and by a vast literature on political culture. In theories of political modernization, increasing popular participation is generally seen as a main characteristic of change (La Palombra and Weiner 1966; Welch 1967, 7f; Finkle and Gable 1968; Huntington 1969; Eisenstadt 1967, 1971; Higgot 1980; Kabashima and White 1986; Leftwich 1990; Linz 1990). Modernization theory (cf. Huntington and Nelson 1977) and theories of political culture (Almond and Verba 1963), of political participation (cf. Milbrath 1965 passim) and of political action (Barnes and Kaase 1979) define structural preconditions or catalysts which favour popular participation in politics: a minimum of economic development and educational standards, as well as certain demographic characteristics with regard to, for instance, race and gender. In the latter approaches, participation as such is mainly treated as a variable dependent on structural determinants and the political system as a framework for action; consequently, empirical research has by and large concentrated on establishing causal relationships between a given individual's political participation and his or her social situation or exposure to politicizing stimuli.

While modernization theories thus concentrate on structural factors, participation and action theories are more interested in positioning an individual within such structures. They focus on psychological (personal) factors, on socialization, and on subjective evaluation (e.g. relative deprivation). On a micro-political level, additional variables such as the responsiveness of authorities to citizens' demands, the coercive potential of the state and ways of justifying both also come into consideration (Barnes and Kaase 1979, 48f).

From a variety of angles, criticism has been levelled against existing approaches to democratic transition processes in ECE (cf. Ágh 1992). But it seems at present unlikely that new conceptual tools will ever be devised. Democratization in ECE is not taking place in a historical or political vacuum, and especially the European Union lends its economic and political support on the basis of a model "European liberal democracy". Moreover, citizens of ECE are also, for the most part, unambiguous about their preference for the European model. It may be argued, therefore, that the study of ECE democratization should depart from existing concepts, test them in a post-socialist context and then proceed to innovative theorizing adjusted to ECE particularities.

Conceptualizing the development of post-socialist parties

On the basis of the theories outlined above, we would reasonably expect political system change in former socialist countries to combine a process of democratic institution building with one of growing participation that includes increasing support of (new) political parties. These expectations have, however, been proved false.

During the early years of transition, most changes took place mainly in the political subsystem. Although it was assumed that economic transition would come more or less automatically with changes in governance and interest group formation, there were even some retrograde tendencies immediately following the first free elections. There were time lags in the development of interest groups and social organizations generally; in Slovenia, for instance, movements had been important actors during the 1980s, but almost ceased to exist towards the end of that decade. Economic and social problems of transition resulted in a drastic decline of mobilization. A political culture of participation and involvement revealed itself as an irretrievable "missing link" at precisely the moment when it was needed most for advancing the process of democratization. It is far from clear whether post-socialist countries can succeed in interweaving developments in their social structure, their political culture and the political system whilst undergoing a continuous "triple transition" (Offe 1991). Although recent evidence suggests that flagging interest in social organizations has been revived, these still cannot be described plausibly as adequate channels of civic participation in political decision-making.

Offe's doubts have, therefore, lost none of their relevance. Under socialism, ECE had undergone industrialization, urbanization, mass education and other aspects of modernization, but had simultaneously frozen modernization in the political sphere for several decades. Whether and how the various speeds of development in these spheres can be harmonized is as yet uncertain. In particular, the following aggravating circumstances confronting post-socialist democratization may be isolated:

- unfavourable economic conditions, which make simultaneous transformation in the economic and political spheres unavoidable (Dahl 1991, 14; Offe 1991);
- a simultaneous distributive and structural transformation of the entire social system, which has important implications for the rebuilding of civil society and legitimate governance;
- difficulties of adjustment to the rule of law;
- problems of institution-building arising from a tendency to apply ready-made solutions to different political and cultural contexts (Zakošek 1992);

- the complexity, shaky legitimacy, fragmentation, instability, and vagueness of party identities in transitional parliamentary systems;
- increasing distrust of political elites;
- poorly developed intermediary channels through which citizens could participate substantially in decision-making (Zakošek 1992, 84f);
- in part resulting from the above, the difficulties encountered in establishing "social partnership" and consultative policy-making;
- problems arising in connection with the replacement of old elites and the emergence of new ones (Weigle and Butterfield 1992);
- and finally, the many obstacles to the development and consolidation of democratic political culture.

Since political parties, at present, dominate the sphere of interest mediation, we must first look into the growth of anti-party sentiment and its structural determinants in post-socialist societies in order to know more about emerging patterns of political participation.

The aversion to parties – a new feature in political participation?

Whatever the differences between particular socialist systems, they all severely curtailed the freedom of association and stalled the development of competition in the political sphere.[1] An apparent, quasi-pluralist diversity of organizations served the interest of the ruling party insofar as it maintained a precarious balance between mobilization and organization and ensured the party's central mediating position (Huntington 1967, 244). But the crucial innovation of this organizational pluralism was perhaps to give oppositional parties at least some scope for legal activity.

Three main patterns, or rather degrees, of political participation in socialist regimes may be distinguished: totalitarian (as in Czechoslovakia), self-managing (as in the weakly totalitarian participatory democracy of the former Yugoslavia) and quasi-liberal (as in Hungary, where this entailed a *de facto* right not to participate, the so-called anti-politics). It will seem obvious from this choice that in ECE, popular notions of politics were clouded by an image of politics as a sinister activity both dangerous and of no interest to ordinary people. Anti-party sentiments were thus an integral component of old regime politics; and if we are to understand the current disenchantment with, and declining interest in, politics soon after the beginnings of democratic transformation, we will surely need to find out how it relates to these earlier experiences.

Symptoms of anti-party sentiment in post-socialism

Is electoral participation declining?

Participation in elections and other forums of more direct democratic participation is perhaps the single most important indicator of a political system's legitimacy. Participation varies, however, significantly even among established democracies.[2]

In Slovenia, participation remained high and stable (82% to 97% between 1989 and 1992), which indicates that voters attributed a high degree of legitimacy to the old regime, to the political system during the most dramatic periods of transformation and to the new democratic system in its initial stages. The process of attaining national independence and creating an independent state partly overlapped with the transformation of the political system, and contributed significantly to high electoral participation (see Table 1). The latter was probably not so much a result of old habits formed under socialism, as the desire for such political changes which would lead to the dissolution of Yugoslavia and Slovenia's sovereignty as a nation-state.

Table 1 Turnout at parliamentary, presidential and local elections and plebiscite in Slovenia in the period from 1986 to 1992 (in %)

	1986	1989	1990a	1990b	1990c	1992
from	82.0	89.7	83.5	76.9	93.5	85.8
to	97.9					

Note: 1986: election of delegates to socio-political chambers in municipal assemblies (percentages indicate highest and lowest turnout)
1989: first round of elections of the president and members of Yugoslavia's Presidency
1990a: first round of elections for the presidency of the Republic of Slovenia
1990b: popular referendum on independence, 23.12. 1990
1992: parliamentary and presidential elections, 6.12. 1992

Source: Election Commission of the Republic of Slovenia

As regards turnout at elections, Slovenia showed marked differences to other post-socialist countries. A significantly higher level of participation in the first free elections may be taken to express the determination to participate in the construction of a new democratic order. But that enthusiasm was somewhat short-lived, and participation has again been a reliably indicator of this development. In Slovenia as elsewhere, after the first free elections voters quickly lost confidence in the new political elites and system, and retreated into passivity. Only recently, apathy seems to have been replaced by the urge to outvote governing parties.

Participation in Poland's parliamentary elections of 1989 amounted to 61.1% in the first round, and to only 25.9% in the second round. Local elections in May 1990 attracted 42.27% of eligible voters, while during presidential elections of the same year, 60% and 53.4%, in the first and second rounds respectively, cast their ballots (White 1992, 155, 157; Batt 1991, 120f). At early parliamentary elections in September 1993, turnout came to 52.08% (Mojsiewicz 1993, 1285). In Hungary, participation in the referendum of 1989 reached 58%, and 63.2% and 45% in the first and second rounds of parliamentary elections in the following year. It fell to 40.18% and 28.4% in the first and second rounds of local elections also held in 1990. Election results (first round) for the municipal assembly in Budapest were even annulled because of insufficient turnout, while the second round still failed to attract more then 35.4% of voters (White 1991, 119ff; Batt 1991, 122). In stark contrast, participation in elections for the federal assembly of Czechoslovakia held in June 1990 reached 96% among Czechs and 94.41% among Slovaks. Though lower at local elections of the same year, 73.5% of Czech and

63.7% of Slovak voters still compare favourably with some of the above results (White 1991, 37, 41; Batt 1991, 126ff). Participation in the former German Democratic Republic was similarly high: 93.4% took part in elections for the GDR *Volkskammer* of March 1990, while 75% cast their ballots in local elections of the same year (White 1991, 75ff). Bulgarian election turnout has also been relatively high, although it has declined recently. At parliamentary elections in June 1990, turnout came to 91%, again at parliamentary elections in the following year it still reached 84%, and at presidential elections in January 1992, 75.41% (Karasimeneonov 1992, 29ff).

We may thus say that in post-socialist countries, participation at elections indicates a rapid decline of public trust in political institutions. Opinion polls concerning the issue of system support confirm this assumption (cf. Illonszki 1990, 16f and Toš 1993). Outvoting incumbent parties, however, seems to be sufficient motivation for marked increases in participation.

Declining party identification

In analyzing party identification in post-socialist countries, it may be useful to distinguish between individuals who show no affinity to any party whatsoever, those who vote consistently for a particular party without being members, and those who adhere to parties as members. In brief, the decline of party identification may be described as the growth of the first group and concomitant shrinking of the other two, and especially the third.

New political systems in ECE are faced with the apparent paradox of drawing on even less support than old regime communist parties. In Slovenia, for instance, between 1970 and 1987 a steady 11% to 14% were Party members. Even during the final years of the League of Communists, when its legitimacy and monopoly had already been seriously undermined, 8.2% of Slovenes upheld their membership. During the early stages of party pluralism, this percentage rose again slightly to 9.1%, but has been declining steadily since. In December 1991, 6.1% of respondents declared themselves party members, and only 4.5% eleven months later. If the degree to which members were involved in party life is taken into account, that decline seems even more drastic. Hungary saw a similar development, with party membership dropping from 8.1% during the final stages of the socialist system to only a few percent at the end of the 1990s (Illonski 1990, 7, 19; Rose 1992, 103; for other ECE countries, cf. Niedermayer and Stoess 1993, 219).

Skepticism towards parties is thus expressed mainly in the small number of people declaring close affinities to any party. In percentages, this share ranges from 3% to 5% in Hungary, the former Czechoslovakia and Poland (Rose 1992, 103), to between 12% and 27% in Bulgaria and Romania. While these differences clearly mirror other divergences in ECE democratization, more research on structural conditions and heritages of political culture will be needed before any definite conlusions can be drawn.

A widening gap between rulers and the ruled

The new parties of ECE are elite parties (Ágh 1992, 19). Party elites can draw on exceptional resources of power appropriated amidst transition euphoria. Once firmly placed, they could, and still can, act without much interference from civil society or variously organized interests, which have disappeared at the end of the 1980s. This means that rulers are at present able to draw up the rules by which they will play, and which will inevitably favour them in future competition. Finally, the power of the elites also reaches deep into other social spheres: senior management, private entrepreneurs, and even the newly emerging interest groups are often dependent on them, and therefore constrained in their interest articulation.

Growing criticism of rampant egotism among political elites therefore seems well-grounded. Opinion polls have also confirmed dissatisfaction with politicians who seem to care mainly for MP's salaries and advancement of their associates, as, for instance, in Slovakia (Miahlikova 1993, 8). 89% of respondents agreed with this perception, while 79% thought that nepotism and careerism prevail in politics, and that good personal connections are indispensable for any kind of career. Mihalikova (1993) perhaps aptly summarizes such views in the adage that "the rich buy democracy; they always have done and they always will".

Citizens of post-socialist countries seem to feel, in the main, helplessly lost and marginalized in the field of political manoeuvrings. Slovenes' responses to the question "Do ordinary citizens in our society today have enough opportunities to influence important decisions regarding social problems?" were similar to those given at the peak of the old regime's legitimacy crisis and just after major system changes (Table 2).

Table 2 He or she has all the possible, or sufficient, opportunities to influence decisions concerning important social problems (affirmative answers in %)

Survey	in percent
SPO 1983	34.6
SPO 1984	27.4
SPO 1986	32.2
SPO 1990	19.9
SPO 1990(2)	30.8

Source: Niko Toš, Slovene Public Opinion research project – SPO (1968–1993), Centre for Public Opinion and Mass Communication, Research Institute for Social Sciences, Faculty of Social Sciences, Ljubljana

Answers show that Slovenes have, of late, felt more marginalized than ever. 71.2% say that they have no influence on politics at all (SPO November 1993), 80.7% feel that politics is merely a struggle for power, a sphere of conflict between competing particularisms (69.1%) or a dirty business (60%) and only to a minor degree an activity aiming at the common good or fundamental values (27.9%).

Such perceptions of politics reveal a profound aversion towards civic activism in general, and encourage the retreat to privacy. Only 19% of Slovaks, for instance, consider substantial involvement in politics a desirable option (Mihalikova 1993, 10) a trend corresponding with interest taken in, and informed opinion on, political matters: in June 1993 only 6% of respondents declared themselves very much interested and very well informed, 24% fairly interested and well informed, 30% as having only a minimum of interest, and 11% as totally uninterested (Mihalikova 1993, 11).

We may thus conclude on the basis of recent opinion surveys as well as other research that all post-socialist countries are experiencing declining levels of confidence in politics and political institutions (for Slovenia, see Table 3).

Table 3 Trust in political institutions in Slovenia (in %)

	totally / a lot	little / not at all	do not know
Trade unions			
SPO1991(1)	15.2	70.2	14.6
SPO1992(2)	16.2	67.8	16.0
SPO1992(3)	13.1	66.9	20.0
SPO1993(1)	11.3	71.3	17.4
SPO1994(4)	12.0	71.9	17.1
Political parties			
SPO1991(1)	12.2	76.9	10.9
SPO1992(2)	11.3	78.3	10.4
SPO1992(3)	7.7	77.2	15.1
SPO1993(1)	3.2	86.6	10.2
SPO1994(4)	4.7	85.1	10.3
Government			
SPO1991(1)	42.6	52.2	5.2
SPO1992(2)	32.5	61.3	6.2
SPO1992(3)	34.3	59.1	6.6
SPO1993(1)	13.3	79.0	7.7
SPO1994(4)	14.3	79.4	6.3
Parliament			
SPO1991(1)	36.8	52.4	10.8
SPO1992(2)	33.7	54.5	11.8
SPO1992(3)	20.2	64.3	15.5
SPO1993(1)	15.0	71.7	13.3
SPO1994(4)	15.0	71.2	13.7
President of Republic			
SPO1991(2)	67.8	26.2	6.0
SPO1992(3)	59.2	32.8	8.0
SPO1993(1)	46.6	47.3	6.1
SPO1994(4)	45.3	46.7	8.1

Source: Niko Toš, Slovene public opinion (SPO), longitudinal project, Centre for Public Opinion and Mass Communication, Research Institute for Social Sciences, Faculty of Social Sciences, Ljubljana

But we should beware of directly comparing these data with similar data for traditional democracies, not only because of discontinuous developments, but because we are, as yet, not in a position to know what exactly we are comparing when juxtaposing responses given in different national contexts.

Parties and their failure to represent social interests

Post-socialist parties were mainly constituted on the basis of anti-communist agendas. They have, in the main, developed in three stages:

- A period of sudden proliferation of opposition parties;
- increasing polarization between transformed communist political organizations and a bloc of opposition parties;
- a renewed dynamics of fragmentation or disintegration of opposition blocs into smaller parties.

But these structural changes in the political sphere have not paralleled transformations in the social sphere. Major social cleavages resulting from economic transition are not reflected in party agendas and policies. Instead, political organizations still concentrate on the ideological distinction between (former) communist and their opponents. Their disregard for changing priorities in society is then, of course, reflected in the public perception of parties as well as in the responses to their actions.

The mushrooming of parties and subsequent fragmentation of parliaments[3] did not entail greater responsiveness to societies' needs and aspirations. Largely incompatible ideological and political structures of constituencies on the one hand and the political sphere on the other were reflected in the discrepancies between party manifestos and party performance, as well as in that between manifestos and voters' preferences. A relatively strong leftist orientation among voters created a vacuum in the political sphere, since even newly emerging leftist parties seemed to have abandoned this faith. Poland's Solidarity leadership, for instance, created such a gap between the organization and its constituency that even some members voted for other candidates (Mojsiewizc 1993, 1288). The most telling line of criticism in this regard was the perception that the elite had betrayed workers' ideals and interests. While the elites were increasingly moving towards the right, voters showed more reluctance to abandon their socialist or social-democratic set of values. Throughout post-socialist societies, social-democratic sentiment was underrepresented in parliaments and the political sphere generally (Szelenyi and Szelenyi 1991, 129ff). It is this representational gap which largely accounts for the recent return of former communists to power. But in Poland, for instance, the process had already begun in 1993, when rightist pary elites were replaced by the reformist Left during early elections in 1993.

Disappointment with party elites and their political orientation has already begun to undermine the legitimacy of new democratic institutions in general, and of parliaments in particular. Since the latter are also the spheres on which post-socialist political life is concentrated, their de-legitimation may well have serious consequences for other spheres (cf. Jackievicz 1992, 18).

An increasing number of people declare that their living standards have declined since 1989, that poverty is spreading while a small minority reap the benefits of transition. In the resulting climate of apathy, early elections seem to have evolved into a sort of plebiscite strategy of simultaneously involving citizens in, and distancing them from, politics (Mojsiewicz 1993, 1285ff; Szelenyi and Szelenyi 1991, 129).

Unconventional political activity as a threat to democratic institutions

Most citizens of post-socialist countries identify democratic transition with a "better life". Against this backdrop of unrealistic expectations, the apparent short term consequences of transformation – privatization, market economy, (sectorially high) unemployment, and accrued difficulties resulting from the general rapprochement between ECE and EU – will provide a fertile ground for increasing, and increasingly intense, social and political conflicts. Already, social problems are the prime motivation for participating in public protest activities, and will probably remain so for some time to come.

In Slovenia, according to the Ministry of Domestic Affairs, 48% of protest activities taking place in 1992 pursued any of a variety of "social" causes, while 29% of them aimed at improving environmental and/or community problems, and 23% had explicitly political objectives. Slovenes are most likely to opt for supposedly unconventional forms of protest in cases of social injustice. According to surveys conducted as part of the 1991 Slovenian Public Opinion project, they will join a strike or protest rally if economic policies should result in a considerable decline of living standards (66.4%), if unemployment should begin to affect "masses of people" (68.4%), and if welfare entitlements and other perceived social rights should be curtailed considerably (68.8%). 56.8% would be willing to defend democracy if a ruling party should abuse its power to destroy it, while other and less drastic symptoms of delegitimation would elicit a much weaker response. Perhaps alarmingly, in 1991 as many as 23.3% of Slovenes stated that they would take to the streets if a party they strongly opposed were to win at elections.

Similarly, citizens of other new democracies (Rose 1992, 65) envisage participation in protest for reasons of unemployment (83%), inflation (76%), and political conflict (52%).

Table 4 Probability of participation in a demonstration (answers in %)

	Bulgaria	Czech Republic	Hungary	Poland	Romania	Austria
very likely	10	4	5	8	10	1
may be	16	28	18	18	21	9
not very likely	18	28	24	22	24	29
definitely not	56	40	50	50	44	60
no answer	1	–	2	1	–	1

Source: Rose 1992, 103

The increasing number of protest actions and other conflicts are obviously a result of the many problems of transition, and may well pose a threat to the stability of new democratic systems. It seems all the more important, therefore, to develop effective intermediary structures and improve communication between political parties and civil society.

The case of Slovenia may serve to illustrate the current deficits of democratic legitimacy and procedural interest negotiation. The public has only two means of influencing politics: elections and strikes. Unlike citizens of traditional democracies, Slovenes cannot rely on conventional forms of influencing government (Table 5).

Table 5 The most effective means of influencing government decisions in Slovenia – first choice and a sum of the first three choices given in 1993 and 1994 (in %)

	1993		1994	
	I	I+II+III	I	I+II+III
elections	43.5	54.8	43.7	54.0
strikes	10.3	28.6	6.3	18.7
demonstrations	3.7	17.5	1.5	12.2
via political parties	3.0	21.0	5.6	29.5
lobbying MPs or ministers	3.9	24.6	4.8	24.3
citizens' initiatives	1.4	10.0	1.1	10.5
various associations	0.6	8.7	0.4	8.5
letters to newspapers	1.0	14.6	3.3	18.6
people do not have any influence	21.5	21.5	14.0	14.0
no answer, do not know	11.1	11.4	19.3	19.3

Source: SPO, November 1993 and November 1994

Even if political parties in ECE try to close the gaps between themselves and their constituencies, it is obvious that they will never be able to substitute interest groups, civic initiatives and critical public opinion. But the development of this intermediary sphere of politics has been hampered by economic difficulties as well as by social structures and a political culture ill adjusted to a modern and stable political system. Old patterns of political behaviour – and most notably the idea that the ruled trade obedience to rulers in exchange for a guaranteed minimum living standard – still prevail. Even though these expectations remain largely unfulfilled, people increasingly assume a passive attitude of waiting for whatever is going to happen. The determination to take action is declining in Slovenia and all other post-socialist nations; they are drifting towards a dichotomy of passive and unstructured masses accumulating potential for eruptive social and political conflict on the one hand, and alienated elites pursuing their own cause and increasingly mismanaging latent conflicts on the other.

In this context, it is hardly surprising that even the dissolution of parliament seems like a plausible (36%) or even desirable (19%) scenario for some voters (Rose 1992, 65). Apathy and avoidance of political participation could easily coincide with both a decline of democratic institutions and practices and with spontaneous and powerful mobilization. The politics of "silent majorities" would then principally be concerned with either welfare issues – i.e. drawing on residual social-democratic values – or those of "law and order – thus appealing to equally widespread conservative sentiment. As a consequence, some observers have argued that the future will see either the rise of Peronism or that of Scandinavian-style welfare government (Szelenyi and Szelenyi 1991, 130f)".

Anti-party sentiment and attitudes towards the multi-party system

Anti-party feelings under the old regimes were a combination of ideological aversion to communism and anti-party sentiment expressed by dissidents, as well periodically in massive protest against declining living standards. The old political elites responded to the first with repression and to the second by guaranteeing employment and a minimum of economic security for all. In Yugoslavia, some sociologists (Županov 1970) even discovered a kind of unwritten social contract between workers and the political elite. The latter's recipe for maintaining peace was to combine a psychology of self-management (stimulating among workers and citizens a sense that they had a share in decisions made in the economic and public spheres), state guarantees of a relatively good (egalitarian) living standard for all, and the real possibility of vertical social mobility through education. In addition, this elite would borrow expensive foreign money with the express purpose of buying its legitimacy at home (cf. Bernik 1992; Lazic 1987; Parkin 1971).[4]

People are generally inclined to see more continuity between old and new political systems than transition euphoria would allow for. Many still appreciate a certain degree of adaptation to popular demands in communist policies. The dominant political culture is not vanishing as quickly, and democratic political culture is not taking hold as effectively, as it might appear from the major normative re-evaluations in the political system. Opinion surveys indicate a gradual and hesitant acceptance of party pluralism among so-called ordinary people (for Hungary, see Illonszki 1990, 19; for Slovenia, see Table 6).[5]

New anti-party sentiment has mostly developed out of a feeling that party elites do not sufficiently care for ordinary peoples' living standards and prospects. To describe politicians as egotistic and indifferent towards a large part of the population, or to say that party elites are fighting ideological battles rather than caring about pressing issues, are alternative ways of expressing much the same thing.

So far, ECE citizens still clearly distinguish between criticism of particular parties and attitudes towards the concept of a multi-party system. Although there is a tendency for criticism to become more generally applied to the sphere of party politics, multi-party democracy is still, for most, the best of all possible political systems. They give preference to the new system when compared to the old one, although this is the consequence of hopes in future development rather than present satisfaction (Table 8). Great expectations seem to be a remarkably constant feature of post-socialist political consciousness.

Table 6 Comprehension of social role of ZK – the League of Communists (in %)

ZK should be...	SPO	agrees entirely	agrees almost entirely	doesn't know – undecided	mostly doesn't agree	doesn't agree at all
a) only one of the political parties – each individual should decide freely which party to support	1986	21.4	24.4	28.7	14.3	11.2
	1988	30.8	25.8	27.0	9.8	6.7
	1989	48.8	26.3	18.0	4.7	2.2
b) the only political power in our society	1986	19.1	26.3	25.9	19.2	9.5
	1988	8.4	10.4	26.1	38.0	17.0
	1989	4.9	7.6	19.3	26.7	41.6

Note: SPO 1986 – SPO 1989: Slovene Public Opinion Poll (SPO) Research in the years 1986, 1987, 1988, 1989 – based on a representative sample of the population of Slovenia.

Table 7 Public attitude towards party elites in Slovenia 1991 – 1992

	Parties serve only the interests of their leaders
January 1991	32.1
December 1991	36.1
March 1992	41.4
November 1992*	56.6

*Note: Question wording 1992: "Politicians abuse politics for their own benefits".

Source: Public Opinion Surveys, Centre for Public Opinion Research, Research Institute for Social Sciences at the Faculty of Social Sciences, Ljubljana

Table 8 Percent of positive evaluation of economic and political systems (average percentage for new democracies)

	past	present	future
economy	57	40	84
political system	38	71	91

Note: Surveys were conducted in Czechoslovakia, Hungary, Poland, Rumania, Bulgaria over a period from November 1991 to January 1992

Source: Rose 1992, 48.

In all post-socialist countries, the notion of democracy has been created as, and continues to be, a social ideal: it is synonymous with greater political and individual freedom as well as with the solution of all major economic and social problems. Slovenes, for instance, relate to democracy primarily as prosperity broadly conceived. It is seen to denote political freedom, freedom of speech and freedom of association (75.5% of respondents during an SPO survey on the subject conducted in December 1991), multi-party systems (70.8%), equal rights for women (69.2%), equality before the law (67.6%), economic improvements (59.4%), more jobs and less unemployment (57.8%), tolerance (55.2%), decision-making at municipal and community levels (52.3%), greater social equality (48.8%), freedom of moral and sexual orientation (47.5%), less corruption in economy and politics (45.3%), and state control over banks and large private enterprises (45.1%).

Only recently, public opinion surveys[6] carried out in ECE countries have indicated declining optimism (Mihalikova 1993, 11 and Toš 1993). The conviction that democracy-building is neither as quick nor as smooth a process as may have been imagined is thus spreading. In Slovakia, for instance, in June 1993, 58% out of a representative sample saw that their country's democracy was both weak and constrained, and were uncertain whether the future would bring more democratic, more authoriarian or more dictatorial rule (Mihalikova 1993, 5). It is as yet hard to tell whether this trend indicates growing realism or a spreading despondency that could itself affect the prospects of building democracy.

Structural and transitional determinants of anti-party sentiment

Perhaps contrary to received wisdom, scepticism towards parties is more widespread in countries with better socio-economic indices of development than in the poorest ECE nations. It is even possible, as Table 9 shows, to establish an inverse correlation.

Table 9 Do you feel close to one political party or movement or not? How close do you feel to this party? (% of all respondents)

	Bulgaria	Czech Republic	Hungary	Poland	Romania	Austria
very	27	4	3	5	12	16
somewhat	28	21	9	11	22	36
not very	6	2	4	3	3	11
total party identification	61	27	16	19	37	63
no party identification	39	72	84	81	63	36

Source: Rose 1992, 103

Paradoxically, we can find a higher degree of party identification in countries with the largest share of people unable to live on their regular incomes alone and therefore least satisfied with their living standard (Rose 1992, 84). Romania and Bulgaria are clearly among the poorest nations of ECE, and their chances for rapid economic and political development might be slim. Yet, their citizens accord the highest degree of legitimacy to the ruling parties. This apparent contradiction can be resolved, however, if the low level of political modernization is taken into account. Democratic transition in either country has so far been mainly formal, with communists still by and large in the position of rulers. Inferior levels of urbanization and democratization, as well as correspondingly prevailing traditionalisms in all spheres, favour a political culture of servility. In this light, it is no longer surprising that the weakest civil societies and the most rudimentarily accomplished democratic transition (Chilton 1994, 4) go together with high levels of loyalty shown towards the (communist) parties.

Present experience and future prospects within a post-socialist framework

In post-socialist countries, the notion of democracy has been shaped as an ideal. Citizens of ECE countries therefore expect it to bring greater individual and political freedom as well as the solution of major economic and social problems. Disenchantment has led some to reject democracy as an unrealistic project, since it is unlikely to meet their unrealistic hopes at least in the short term range. This change of heart could in itself have repercussions on the pace and quality of democratic transition, although expectations seem, on the whole, to converge gradually towards a consensus that democratization is a very gradual development.

Obstacles to democratization can be found in the political elite and their practices, which have undermined confidence in leaders and institutions generally, and in the population, which lacks a strong and active middle class usually instrumental in organizing and articulating interests and which shows a tendency of relying on the state to solve problems of immediate concern to all.

As regards prospects for democratization and the development of a democratic political culture, even social scientists may by now be sorted into convinced optimists and pessimists. Some think that retrograde and authoritarian tendencies will prevail, while others are confident that substantial democratization and differentiated interest mediation as well as participation will come about. These alternative scenarios could be, and have been, presented as those between social democracy and Peronist politics. Perhaps Datulescu (1992, 21) is close to a workable solution when he suggests that it might be found in a social democracy implementing conservative policies without appearing to renounce community values. The latter would seem of particular necessity if the gap between the poor on the one hand, and the new entrepreneurial and bureaucratic establishment on the other is not to widen to such an extent that it might further encourage extremist politics. Since these could considerably complicate a variety of problems, national or ethnic ones not least among them, they might well become catalysts for halting the democratization process in some ECE countries (Karasimeonov 1992, 26).

Finally, two important lessons drawn from the above are that the development of post-socialist societies is as yet close to unpredictable, and that party elites will be critical in shaping future democratization one way or another (cf. Pridham 1990, 105; Ágh 1992, 24; Lewis 1993, 306)

Notes

1 Formal guarantees of the right to free association in socialist constitutions were rendered worthless by its subordination to supposedly higher interests of the community, the Communist Party, or whatever. Take, for instance, Article 51 of the Soviet Constitution of 1977, or Article 153 of the Yugoslav Constitutions. See Heller (1986) and Fink-Hafner (1990) for details.

2 i.e. from 50% (USA, Switzerland, Turkey) to 86% and over (Australia, Sweden, Denmark, Italy). See Powell (1982, 15) for detailed information.

3 40 parties registered for the first free elections in Bulgaria, 26 in Czechoslovakia, 35 in the GDR, 75 in Romania and 50 in Hungary. Of the 75 Romanian parties, for instance, only nine gained seats in parliament (Datculescu 1993, 4ff). Indices of fragmentation remained high throughout post-socialist countries, even where entrance thresholds to parliament were raised in order to limit such fragmentation. Slovenia's index was 0.8375 in 1990 and 0.840 with a 2,5% threshold; Hungary's index was 0.731 in 1990 and with a 4% threshold, in the Czech National Council it was 0.550 (5% threshold) in 1990, and in the Slovak National Council 0.799 with a 3% threshold. According to Sartori (1976), all the above except for the Czech National Council belong in the category of extremely fragmented and polarized or semi-polarized democracies (Sartori 1976, 314).

4 Following a protracted economic and political crisis at the end of the 1960s, the Yugoslav government borrowed heavily abroad in order to offer workers cheap long term loans enabling many to build their own homes.

5 In Slovenia, attitudes towards the League of Communists were measured in opinion polls between 1968 and 1988. Although the share of respondents who believed that the government acted in accordance with majority interests declined steadily from between 20% and 50% to 5.8% in 1988, an increasing percentage thought that it acted at least *partially* in accordance with majority interest (44.1% in 1988). Only 18% rejected the communist government's policies whole and entire. Source: Slovenian Public Opinion Research 1968-1988, Centre for Public Opinion, Faculty of Social Sciences, Lubljana.

6 e.g. the Slovenian Public Opinion Survey December 1993, or Slovakia's Public Opinion Survey of June 1993.

References

Ágh, A. (1992), 'The Emerging Party System in East Central Europe', Budapest Papers on Democratic transition, No. 13, Hungarian Centre for Democracy Studies Foundation, University of Economics, Budapest.

Almond, G. A. and Verba, S. (1963). *The Civic Culture*, Princeton University Press, New Jersey.

Almond, G. A. (1968), 'A Developmental Approach to Political Systems', in Finkle, J. L., Gable, R. W., *Political Development and Social Change*, John Wiley and Sons, New York et al., pp. 96–118.

Almond, G. A. (1993), 'The Study of Political Culture', in Berg-Schlosser, D. and Rytlewski, R., *Political Culture in Germany*, Macmillan, London, pp. 13–26.

Barnes, S.H. and Kaase, M. et al. (1979), *Political Action. Mass Participation in Five Western Democracies*, Sage, Beverly Hills and London.

Bates, R. H. (1991), 'The Economics of Transitions to Democracy', *Political Science and Politics*, vol. 24, no. 1, pp. 24–27.

Batt J. (1991), *East Central Europe from Reform to Transformation*, The Royal Institute of International Affairs, Pinter Publishers, London.

Bernik, I. (992), Dominacija in konsenz v socialistiňdružbi, Znanstvena knjiznica FDV, Ljubljana, No.2.

Berglund, S. and Dellenbrant, J. A. eds. (1991), *The New Democracies in Eastern Europe. Party Systems and Political Cleavages*, Edward Elgar, Aldershot et al.

Chilton, P. (1994), 'Mechanics of Change: Social Movements, Transnational Coalitions, and the Transformation Processes in Eastern Europe', *Democratization*, 1994.

Dahl, R. A. (1971), *Polyarchy. Participation and Opposition*, Yale University Press, New Haven and London.

Dahl, R. A. (1991), 'Transitions to Democracy', in Szoboszlai G. (ed), *Democracy and Political Transformation. Theories and East-Central European Realities*, Hungarian Political Science Association, Budapest, pp. 9–21.

Datculescu, P. (1992), 'The New and the old Parties of the Nineties in Romania – in Search for Potential Supporters', paper presented at an International Conference on Political Parties in the New Europe, Vienna, 24–26 April.

Eis, Z. (1992), 'Tradition und Gegenwart der politischen Kultur der Tschechoslowakei', in *Zeitschrift für Politik*, yr. 39, vol. 2, pp. 161–170.

Eisenstadt, S. N. (1971), 'Patterns of Political Modernization', in Welch (1967) pp.246–266.

Eisenstadt, S. N. (1971), 'The Scope and Development of Political Sociology: Political Modernization and Political Sociology of the Modern State', in Eisenstadt, S. N. (ed.), *Political Sociology*, New York and London, pp. 317–338.

Feher, F., Heller, A. and Đerd, M. (1986), *Diktaturea nad potrebana*, Rad, Beograd.

Fink Hafner, D. (1992), 'Political Modernization in Slovenia in the 1980s and the Early 1990s', in *The Journal of Communist Studies*, vol. 8, no. 4, pp. 210–226.

Fink-Hafner, D. (1993), 'Socio-Economic Transformation – The Only Missing Link in Democratisation of a Policy Process in Post-socialist countries?', paper written at EPPI, University of Warwick.

Fink-Hafner, D. and Svasand, L. (1994), 'The Status of Political Parties in Old and New Democracies', in *Small States in the New European Architecture. Slovenia and Norway*, Department of Comparative Politics, University of Bergen, Bergen.

Finkle J. L., Gable R. W. (1968), *Political Development and Social Change*, John Wiley and Sons, New York.

Gabanyi, A. U. (1992), 'Rumänien zwischen Revolution und Restauration', Aus Politik und Zeitgeschichte, Beilage zur Wochenzeitung Das Parlament, Bonn and Trier, 14/92, 27 March, pp. 23–31.

Higgott R. A. (1980), 'From Modernization Theory to Public Policy: Continuity and Change in the Political Science of Development', *Studies in Comparative International Development*, vol. XV, no. 4.

Huntington, S. (1967), 'Political Development and Decay', in Welch (1967, 207–246).

Huntington, S. (1969), *Political Order in Changing Societies*, Yale University Press, New Haven.

Huntington, S P. and Nelson, J. M. (1977), *No Easy Choice. Political Participation in Developing Countries*, Harvard University Press, Cambridge, MA and London.

Ilonszki, G. (1990), 'Political Participation in Hungary', *Strathclyde Papers on Government and Politics*, no. 76, Department of Government, University of Strathclyde, Glasgow.

Jackiewicz (1992), 'The Roles of Polish Parlamentarians in the Post-Communist Years', *Budapest Papers on Democratic Transition*, Department of Political Science, Budapest University of Economics, no. 27.

Kabashima, I. and White, L.T. III (1986), *Political System and Change*, Princeton University Press, Princeton, New Jersey.

Karasimeonov, G. (1992), 'The Post-Communist Party Panorama. The Case of Bulgaria', paper presented at the International Conference on Political Parties in the New Europe, Vienna, 24–26 April.

Karasimeonov, G. (1992a), 'Vom Kommunismus zur Demokratie in Bulgarien', *Aus Politik und Zeitgeschichte, Beilage zur Wochenzeitung Das Parlament*, 27 March, pp. 13–22.

Kitschelt, H. (1992), 'The Formation of Party Systems in East Central Europe', in *Politics and Society*, vol.20, no. 1, March, pp. 7–50.

Lazic, M. (1987), *U susret zatvorenom drustvu? Klasna reprodukcija u socijalizmu*, Naprijed, Zagreb

LaPalombara, J. and Weiner M. (eds.) (1966), *Political Parties and Political Development*, Princeton University Press, Princeton, New Jersey.

Leftwich, A. (ed.) (1990), *New Developments in Political Science*, Edward Elgar, Aldershot et al.

Lewis, P. G. (1993), 'Democracy and Its Future in Eastern Europe', in Held, D. (ed.), *Prospects for Democracy, North, South, East, West*, Polity Press, Cambridge, pp. 290–311.

Linz J. J. (1990), 'Transitions to Democracy', *The Washington Quarterly*, vol. 13, no. 3, pp. 143–164.

Lipset, S. M. (1959), 'Some Social Requisites of Democracy: Economic Development and Political legitimacy', *The American Political Science Review*, vol. 52, pp. 69–105.

Markič, B. (1993), Državljan in demokracija, *Teorija in praksa*, vol. 30, no. 1–2, pp. 85–91.

Mihalikova, S. (1993), 'The End of Illusions: Democracy in Post-Communist Slovakia', paper presented at the Seminar on Constitutionalism and Politics, Piestany, 11–14 November.

Milbrath, L. W. (1965), *Political Participation. How and Why Do People Get Involved in Politics?*, Rand McNally and Co., Chicago.

Mojsiewicz, C. (1993), 'Parlamentarne volitve na Poljskem', *Teorija in praksa*, Ljubljana, vol. 30, no. 11–12, pp. 1285–1290.

Niedermayer, O. and Stoess, R. (eds.) (1993), *Stand und Perspektiven der Parteienforschung in Deutschland*, Westdeutscher Verlag, Opladen.

Offe, C. (1991), 'Capitalism by Democratic Design? Democratic Theory Facing the Triple Transition in East Central Europe', *Social Research*, vol. 58, no. 4, pp. 864–902.

Parkin, F. (1971), *Class Inequality and Political Order. Social Stratification in Capitalist and Communist Societies*, Praeger Publishers, New York and London.

Pye, L.W. 'Political Culture', *International Encyclopaedia of Social Sciences*, pp. 218–225.

Powell G. Bingham, Jr. (1982), *Contemporary Democracies. Participation, Stability, and Violence*, Harvard University Press, Cambridge, MA and London.

Pribersky, A., Kovacs, E. and Kiss, J. (1992), 'Das Mehrparteiensystem in Ungarn', Dr. Karl Renner Institut, Publication no. 14, Vienna.

Pridham, G. (1990), 'Political Actors, Linkages and Interactions: Democratic Consolidation in Southern Europe', *West European Politics*, vol. 13, no. 4, pp. 103–117.

Rose, R. (1991), 'Between State and Market. Key Indicators of Transition in Eastern Europe', *Studies in Public Policies*, no. 196.

Rose, R. (1992), New Democracies Between State and Market. A Baseline Public Opinion Report, *Studies in Public Policies*, no. 204.

Roskin, M. G. (1994), 'The Emerging Party Systems of Central and Eastern Europe', in *Democracy in the 1990's – A Special Issue of Global Issues in Transition*, no. 6, pp. 62–76.

Sartori, G. (1976), *Parties and Party Systems. A Framework for Analysis*, Cambridge University Press, Cambridge.

Szelenyi, I. and Szelenyi, S. (991), 'The Vacuum in Hungarian Politics: Classes and Parties', *New Left Review*, no. 187, pp. 121–137.

Taylor, C. (1992), 'Parties in Search of Cleavage. Elite-Mass Linkages in Hungary', Budapest Papers on Democratic Transition, Hungarian Centre for Democracy Studies Foundation, Department of Political Science, Budapest University of Economics, no. 16.

Toš, N. (1993), *Slovensko javno menje, raziskovalni projekt* (Slovene Public Opinion research project), Ljubljana, Centre for Public Opinion Research, Faculty of Social Sciences.

Weigle, M. A. and Butterfield, J. (1992), 'Civil Society in Reforming Communist Regimes: The Logic of Emergence', *Comparative Politics*, vol. 25, no. 1, pp. 1–23.

White, St. (ed.) (1991), *Handbook of Reconstruction in Eastern Europe and Soviet Union*, Longman, London.

Zakošek, N. (1992), 'Choosing Political Institutions in Post Socialism and the Formation of the Croatian Political System', *Croatian Political Science Review*, vol. 1, no. 1, pp. 79–90.

Zupanov, J. (1970), 'Egalitarizam i industrijalizam', *Sociologija*, Beograd, vol. 12, no. 1, pp. 4–45.

Welch, C. E. Jr. (ed.). (1967), *Political Modernization. A Reader in Comparative Political Change*, Wadsworth Publishing Company, Belmont, CA.

6 Political Culture in Slovenia

Igor Lukšič

Some observations on the general concept

Political culture was invented as a concept for the evaluation of different political systems[1], and as a tool for thinking both the cold war and decolonization in the context of an emerging North American post-war hegemony. It was intended as a variable supposedly explaining why the Anglo-American political system is superior to any other.[2] Once political culture had achieved hegemonic status in political science,[3] many scholars tried to describe the differences between (American) civic culture and the political culture of their own country in terms of gaps or shortcomings of the latter. They also advanced proposals for changes in the political culture of their countries which would make them more civic and thus more adapted to (the American perception of) democracy. The so-called gap between *praxis* or activity on the one hand, and values, awareness or consciousness of political activity on the other, has been modelled on a Newtonian paradigm[4] of the perception of basic realities fundamental only to European culture and its understanding of common sense. The gap appears only from the viewpoint of an outward (scientific) observer, but not from the perspective of those sharing and involved in a particular (political) culture. It is, moreover, a gap which exists only for one type of observation,[5] namely that which cannot envisage practice and consciousness as inseparable.[6]

During the second half of the 1960s, we witnessed a "revolution in comparative politics" (Verba 1967, 111) as well as two studies on political culture which departed from the mainstream. In an historical analysis written in German, Lehmbruch (1967), examined the political cultures of Switzerland and Austria. He described the creation and functioning of two political cultures embodying the principle of *amicabilis compositio* (amicable agreement), and develops a concept of *Proporzdemokratie* and/or *Konkordanzdemokratie*, which seems to have been sufficiently un- (though not anti-) American and remote from the hegemonic perception to be relegated to the margins of the debate.[7]

Lijphart (1968), writing in English and in a distinctly American vein, approached political culture in the Netherlands with the concept of "politics of accommodation", presented as the essence of Dutch political culture. In a variation on Almond's classificatory scheme, he adds the dimension of elite political culture: where the political culture of the masses is fractioned, the political elite can stabilize democracy by means of a politics of coalition rather than competition. In a consociational democracy, representatives of all important segments or political cultures form a coalition while retaining their autonomy, and based on principles of proportionality and mutual veto. In this way, Lijphart explains the stablility of such diverse democracies as Switzerland, Austria, the Netherlands and Belgium, as well as other political systems.[8] A huge consociational school applying Lijphart's concept to Northern Ireland, South Africa, Yugoslavia, Italy, Canada, Uruguay, Israel, Lebanon, Norway and other countries has emerged.

Both Lehmbruch and Lijphart have thus contributed to the criticism of Anglo-American notions of the right kind of political culture for the right kind of democratic system – a criticism vindicated by a succession of anti-war, student, anti-racism and other protest as well as by the political consequences of Watergate. At the same time, they have called attention to the fact that political culture can be homogeneous only on one level, while it is differentiated or even organised into subcultures on others.

Since the political systems of East Central Europe have changed labels, the concept of political culture has been used as a tool for the Americanisation of communist savages. Or, in the words of Gabriel Almond (1993, 14), "conceptual jargon, like passwords in warfare, often served the purpose of defining friend and enemy, rather than enhancing our capacity to explain important things". But there is another largely unexplored option – to conceive of political culture not as an ideological project, but as a branch and category of political anthropology. Within such a framework, political culture can be read only from the viewpoint of political life and behaviour of *zoa politika* among themselves and in interaction with other societies. It is an ensemble of patterns of political behaviour – including patterns that cannot be isolated by any supposedly objective scientific approach – and may be described as the sum of self-perceptions, reflection and actions relating to aims, values and feelings. Perception is the consciousness of such patterns and thus constitutes the subjective aspect of political activity, or the creative and individual act of shaping social relations. Perception also includes rationalization and awareness of that activity – an awareness which in turn feeds into *praxis* as one of its inseparable components.

If we limit the scope of analysis to directed self-perception in the way opinion surveys do, we will discover only value preferences and assessments produced by those surveys. Such research serves mainly as a tool for comparing, and especially ranking, regions or countries, and it is not hard to imagine Pavlov's dogs across Europe and the United States reacting to the sound of bells according to variations in sound colour or wavelength. Researchers commonly start with a clear idea of proper and improper political values, and subsequently apply the hegemonic (proper) ones as measuring rods to determine the relative shortcomings of other, improper, values.[9] But this kind of research makes sense only as a way of strengthening the hegemonic pattern of political culture: the conviction that to have some kind of political opinion is reason enough not be involved in politics any further; or that having an opinion already constitutes political involvement.

The concept of political culture in Yugoslavia and Slovenia

A similar logic has been applied in the introduction of the concept to Yugoslavia in general, and Slovenia in particular, during the late 1960s and early 1970s. Yugoslavia's political leaders recognized that the prevalent political culture was far from adapted to the political system of self-management. In the perception of the political elite, the results of political culture research should have suggested that the practice of self-management had more imminent democratic potential than that of any other country, including the United States; and once these hopes had proven futile, researchers suggested ways of making political culture more self-managemental by, for instance, appointing appropriately oriented individuals to key positions in the political system (Novosel 1969, 63).

The ensuing debates on political culture revealed a great diversity of opinions regarding the content and scope of the term. Some doubted whether political culture could be said to exist in Yugoslavia at all, while others tried to forge a concept of "self-managemental political culture". The most elaborate version of the latter was developed by J. Đorđević (1981), who thought that self-management needed to go beyond political action or judicial regulation and be conceived as an imaginary space and vital nerve of any political system embodying it. The political culture of self-management was to provide an alternative to the *étatiste* culture of a fetishized state on the basis of democratic principles. Out of twenty-two features of a political culture based on self-management, Đorđević's priorities were, in this order, the primacy of civil society over the state, socialization and circumscription of the role of the state, and the safeguarding of pluralism against monist or centralist tendencies. As general characteristics of such a culture he described optimism, revolutionary romanticism, emotional expressivity, anti-monism and an aversion to uniformity. It must be stressed, however, that the political culture of self-management was mainly a normative project legitimizing the introduction of new subjects in secondary school and university curricula, as well as in the training of delegates and leaderships of sociopolitical organizations. In Slovenia, the concept of political culture of self-management found its only echo in an article by Rojc (1975, 35) arguing that "school should be the creator of a broad self-managemental culture in the personality of every manager (i.e. working man and woman) ".

In Slovenia, reception of political culture remained limited to the academic sphere. Since 1968, data have been collected in the form of annual public opinion surveys conducted by the Institute of Public Opinion and Mass Media Research at the Faculty of Social Sciences, Ljubljana.

The first major study of political culture in Slovenia was published by S. Južnič's (1973). Using an interdisciplinary approach to political culture, he addressed the relationships between individual and society or culture, as well as between political and other forms of socialisation. Južnič also examined agents of socialization and ways of classifying political cultures. Political culture, for him, is an anthropological category rather than an attribute of individuals lending itself to measurement by public opinion surveys. In 1989, a revised and enlarged version appeared in print which included essays on political manipulation, biological determination of political behaviour, and on political language and the ritualisation of political power. Južnič's conceptual tools have been the most important stimulus of research in this field, even though his empirical forays into Slovenian and Yugoslavian political culture are scarce. By now, of course, many more studies on the political culture of Slovenia have been carried out in other fields.[10]

Political culture in Slovenia

If we read political culture from within the political practices presently found on Slovenian territory, it will be readily apparent that the country is governed by corporatism, and democratic only in the sense that democracy today covers all spheres of politics and everyone believes in Democracy. Slovenia's political system is by no means liberal, though it may well be freer than that of, say, the United States or Britain, since there has at least never been an opportunity to oppress other nations or races. But Slovenia is also a pluralist society (Furnivall, 1939) in which many political cultures coexist. They may be described as the three pillars of Catholicism – the most organized – socialism and liberalism. The first has a predominantly rural base, while towns are the main strongholds for the other two. Corporatism is strongest in the Catholic and socialist pillars, and still remarkably prevalent in liberalism when compared to European equivalents.

Slovenian corporatism as a principle of social organisation is based on the metaphor[11] of the *living body*. It is an organicist conception granting the three pillars of corporatist structure a status of vital interests of society and deriving functional interests as well as funtional representation from them. Society, on this view, is a living being whose survival depends on the health of all its organs. State, politics and society are inseparable; the paramount concern of politics are the vital interests or pillars, while the aspirations of other groups or individuals must necessarily submit to the former. Accordingly, corporatism's most cherished values are a stable social, national, political, communal and familial order, as well as unity, solidarity, justice, and equality. It is, in other words, one marked by tendencies towards nationalism, patriarchalism and authoritarianism.

Interest representation takes place through a complex network of special interest and professional organisations (or, perhaps better, communities), which are incorporated in the political system. Parties are not recognized as adequate mediators of interests, since politicization is seen as a distortion of "real" social interests. Anti-party attitudes and anti-parliamentariansm are thus also features of the corporatist perception of politics. Social pacts between industrial and rural labour, business and other allegedly vital organisations of society are favoured. Formal procedures in politics are regarded as less relevant than its content, and the "substance" of interest articulation is seen as more important than the modes of such articulation. In short, legitimacy always takes precedence over legality, and the politics of negotiation and reconciliation in Slovenia are dominated by moral and religious attitudes rather than rationalist or legalist approaches.

Catholic corporatism

Perhaps the most important pattern of behaviour and perception of politics was brought to Slovenia with Christianization, beginning in the 8th century and reinforced during the Counter-Reformation of the early 17th century. The Reformation had provided a basis for freedom of thought by allowing nobles and citizens a choice of religious service; it had created Slovenia's literature and standardized its language. It had introduced the vernacular in churches and in new schools for the people. In sum, the Reformation may be described as a period of cultural, proto-national, political and economic Slovenian revival. Its peasant riots set a precedent for every later revolt against exploitation, oppression and occupation.

But then came the Catholic response, its book burnings, dissolution of non-Catholic communities and expulsion of a large part of the nobility and the educated. Witch hunts[12] and autos-da-fe strengthened re-Catholicization on Slovenian territory. Whereas Europe, following the ferocious wars of the 17th century, began to develop a pattern of religious toleration which would eventually provide a basis for the pluralism created at the beginning of the 20th, Slovenia produced a Catholic corporatism grounded on the assumption that Slovenes were to be Catholics or else not to exist. Catholic political culture in Slovenia was in its origin exclusionary and intolerant, and fostered solidarity by means of such exclusion.

Slovenia never had an influential aristocracy of its own, and its citizenry or bourgeoisie was as weak as it was lacking in assertiveness. While Slovenes counted mainly as soldiers and taxpayers, there were no strata representing a general will or interest of the state. Since the state therefore provided no framework for the construction of a political identity, the realm of the common became the domain of the Catholic church. From the beginning of the 17th to the end of the 19th century, it was an unrivalled, and perhaps even the only, sphere of the public; it was accordingly defended by all possible means. Slovenia has also been, in the main, a nation of peasants[13] and Catholics until relatively recently, and has developed localism as a core element of its political culture; for centuries, Slovene peasants lived village lives and were open neither to neighbouring villagers nor to the world at large.[14] Moral, political and cultural boundaries coincided with those of local communities. In Slovenian ideas about politics, only the real and concrete person with equally real and concrete interests as well as obligations towards the community exists. The absence of a state and a market economy brought with it the impossibility of conceiving an abstract public space and citizenry, and therefore also the impossibility of civic equality and individual liberty.

Under these circumstances, only a culture limited to governing one village, one farm, one family and, of course, one Catholic community could flourish. Following the *Encyclica Rerum Novarum* of Pope Leo XIII (1891), a powerful peasant movement named "Krekovstvo" after its leader, Janez E. Krek, emerged (cf. Žižek 1987, 9–60 and *Revija 2000*, nos. 41–42, 1989). Its main goals were the protection of Slovenes and their language, the regrouping of peasants in cooperatives, their instruction in new farming techniques and help during crop failures. Promoting corporatist ideals of a Christian community, and of national unity, Krekovstvo was primarily anti-liberal and anti-capitalist in orientation.

In the 1930s, following the *Encyclica Quadragesimo Anno* of Pius XI, Christian intellectuals promoted the idea of the corporatist state. Corporations were construed as marking organic boundaries between the individual and the state, as well as between "lower" and "higher" orders of society. Andrej Gosar, one of Slovenia's most influential political thinkers, developed a radical critique of parliamentary democracy and suggested replacing it by a system of popular self-government and socialization of the economy. Gosar represented a, relatively speaking, modernist current within corporatist thought, besides which existed the authoritarian corporatism of Ivan Ahčin, Jože Jeraj or Jakob Aleksič, and the radical corporatism of Ciril Žebot all but indistinguishable from fascism (Zver, 1992).

In the Kingdom of Yugoslavia, Slovenia's political sphere was created and monopolized by the Catholic Peoples Party, while the Socialist movement was a target of continuous harassement and repression. Its radical wing, the Communist Party, was outlawed beginning in 1921. Liberals, on the other hand, were mainly oriented towards Belgrade and devoid of any real influence on the Slovenian political scene.

Socialist corporatism

After its suppression in the 16th and 17th centuries, the Slovenian reformation was in some sense revived by labour movements of the late 19th century. Socialist intellectuals and a Social Democratic party at last provided for some competition in the Catholic-dominated public sphere, and their appearance on the political scene inevitably entailed head-on clashes with the Church. The Slovenian state and its capitalists were not the only, not even the main, foes of socialism. The Church was the largest landowner as well as the preponderant ideological and social power. It had also to a large extent prevented the rise of a strong citizenry and the development of a modern state. Slovenian socialism, therefore, had to define its role less as that of an undermining force than as that of an agent of state-building.

Whenever political activists or writers have tried to rally support against authoritarianism and the suppression of freedom, they have begun by acknowledging that Slovenes are a nation of *Knechte* or servants.[15] This was most obvious in the attempt to organize resistance against Nazi occupation and assimilation, for which the slogan of "active Slovenehood" – as against patterns of monolithic thinking and servility – was coined. Soon it became clear to all concerned that in mobilizing the Slovenian masses, the Liberation Front would change Slovenian national character, and that the country's political culture would have to be transformed if a modern nation state was to be created. Unsurprisingly, the Church showed little intention to support this project, since to construct such a state meant first and foremost to unhinge the Church as a near-absolute political and societal power.

Besides having been instrumental in bringing about the movement for national liberation, socialism was also the first social and political force since the Reformation which offered a strongly coherent set of ideology and practice opposed or alternative to Catholicism. The struggle against fascism was at the same time a struggle for tolerance and new pattern of coexistence between Catholicism and socialism. But the upshot of these endeavours was a rehearsal of an older one: socialists and nationalists on the one hand, and Catholics and occupiers on the other, were divided by an ideological and political frontline, each with their partisan (or collaboration) army.

During World War II, the foundations for a new state were laid. Following the defeat of fascism and National Socialism, the Catholic church lost much support due to its close ties with both, and was removed from the pedestal of Slovene political history. In the context of the reconstructed federal Yugoslavia, the struggle between the Church and secular social forces ended, of course, with the construction of a socialist, rather than a liberal state. Catholicism's hegemony was supplanted by a socialism functioning also as a kind of secular religion.[16] In the new Yugoslavia, the problem of coexistence was solved by granting the Catholic Church a status and role equal to that of orthodox Christianity and Islam. Nevertheless, in a Slovenian context the Catholic church largely retained its strength as a social pillar or segment, and it built up new structures which included its own print media, associations for all social strata, youth and interest organisations, etc.

Under socialism, the nation state increasingly took hold of Slovenes' political imagination. It had existed previously, but in a much more fragmented and abstract or "spiritual" form very gradually developing through the creation of parties, cultural movements, during two months of independent statehood in 1918, and the subsequent creation of a Southern Slav Kingdom of Serbs, Croats and Slovenes (renamed Yugoslavia in 1929). The constitutions of 1963 and 1974 may be said to represent the pinnacle of Slovene national consciousness prior to the erosion of the socialist regime. Slovenia, like the other republics, already enjoyed a measure of autonomy within the federation. The creation of Slovenia's independent nation state only apparently ended the parallel development of nationalism and federalism. The latter has been reoriented towards membership in a federal Europe and active involvement in international, and thus in some sense globally federal, organizations.

After World War II, the concept of corporatism was further developed under the hegemony of socialist ideology (Lukšič 1992). "Real" interests and their non-politicized representation within the framework of the Party was to be achieved through "pluralism of self-managemental interests" (Kardelj 1977). Such pluralism took it for granted that parties and their contentious interest representation would destabilize the country and eventually lead to civil war.[17] Instead of a party system, a web of socio-political, professional and interest organisations with the League of Communists[18] acting as an umbrella and playing the leading role was established. The constitution of 1974[19] introduced self-management communities for "vital interests of society": education, culture, social care, medical care, forestry, farming, transport, sport, energy, retirement, child care. Whenever matters falling within their sphere of competence were discussed in Pariament, each of them could participate in the deliberations as a fourth chamber. In addition, a delegate system was to ensure the voicing of concrete and pressing concerns. Assemblies were constituted on the basis of "self-management elections" rather than on that of general elections (Kardelj 1977, 109). On the the level of the republic, a three-chamber parliamentary system comprised the Socio-Political Chamber consisting of delegates from various socio-political organizations, the Chamber of Communes and its representatives from all administrative communities, and the Chamber of Associated Labour with representatives from all sectors of the economy and social services. Self-managemental agreements and social compacts between enterprises, organized working people and communities were to replace the allegedly anarchic forces of the market and of state interventionism. Self-managing society was intended to absorb civil society and the state, and was ultimately modelled on idealized visions of village community and common property or social ownership. Yugoslavia's political system was, among many other things, also a *de facto* federation of communes, which were recognized as foundations of the state and which sent delegates to both republic (Chamber of Communes) and federal (Federal Chamber) assemblies. All these forms of political and social activities were supposed to enhance a political culture of participation extending to all "working people". During the times of socialist hegemony, a number of explicitly political institutions were also modelled on corporatism: the Public Front, the Anti-Fascist Front (which included all associations and organisations except the Catholic Church and other fascist collaborators), the federal Chamber of Producers (1953–1963), chambers of functional representation on a variety of levels ranging from the communal to the federal (1963–1974), and institutions of "integrated self-management" (1974–1990 or 1992).

Pluralist corporatism

Pluralism has become the new hegemonic ideology in Slovenia. Parties have been established, but almost all of them resemble corporate institutions rather than organizations with political aims. Parties are supposed to serve funtional interests and not to engage in a struggle for political power. Some pensioners' parties (Grey Panthers, Democratic Party of Retired People), for instance, were set up with such an – implicit – understanding of the role of parties. Similarly, a Liberal Party and a Party of Tradesmen, a Peasants' Party (now renamed Peoples Party), parties of intellectuals and artists (Democratic Alliance), and of labourers (Labour Party as well as two Social-Democratic parties) all espouse a corporate cause. To advance the ideological causes of some Catholics and nationalists, a Christian Democratic and National Party have been created. Situated somewhere between ideological rallying and the representation of functional interests were the Greens, whose ideologist and pragmatist factions eventually broke apart in 1994. Only two parties were political first and foremost: the former League of Communists (renamed the United List of Social Democrats), and the former Socialist Youth organization turned Liberal Democracy of Slovenia.[20] Grounded in Slovenia's history of anti-party attitudes, the country's first step towards pluralization was essentially corporatist in character.

Slovenia's parties had originally developed out of a wealth of social organizations and periodicals devoted to a variety of political causes. Even under single-party rule, a degree of pluralism survived in the form of ideologically motivated journals and reviews,[21] which in 1980s provided a basis for the formation of new political parties. A further step towards politicizing the representation of aims and interests was the renaming of various sub-groups incorporated in the Socialist Alliance up to 1989. But the concept of a purely political party has still not been widely accepted in Slovenia. Only during two very short periods could parties occupy the centre of public attention: shortly after the creation of the transient Slovenian state in 1918, and during the early 1990s, again, of course, in the initial stages of independent statehood. At all other times, as I have been arguing, politics has always seemed as something tainted, remote and filthy to Slovenes.[22] Slovenes have generally only taken to active politics as a last resort, that is, in the form of peasant riots or wars of national liberation, during decisive stages of state building, or when their bare existence was threatened in some form or other. Out of such necessarily transient rebellious phases, no pattern of long-term involvement in politics has ever evolved. Parties were, on the contrary, mostly perceived as sources of conflict and tools of a small, manipulative and corrupt elite caring only for itself.

Pluralization in Slovenia has so far produced 124 parties. But at local elections held in 1994, only thirty-seven of them took part, and only seven of them made their way into representative institutions. In the constitution of 1991, parties feature only in the stipulation that members of the police or armed forces as well as the Constitutional Court may neither join political parties nor compete for party positions. Parties have even been implicitly designated as non-constitutional forms of political activity. Article 82 of the Constitution states that representatives ought to defend the interests of the entire people and "shall not submit to undue influence from any source". The new State Council is a corporatist institution based on principles of functional representation and indirect election from, yet again, "vital interest organizations" such as trade unions, employer's organisations, representatives of universities, agriculture, tradesmen, physicians, nurses, and 22 local communities. Although the Council has very few competences, it nevertheless stands as a powerful symbol of distrust vis-à-vis political parties (cf. Keber 1992) and has provided political traditions with a *de facto* legitimacy and mandate to slow the progress of liberal democracy (Lukšič 1994, 207).

Since individuals joining parties have generally been found unwilling to make regular financial contributions, a law had to be adopted in 1994 providing for their maintenance out of public funds. The climate of hostility or indifference to parties is thus forcing the political elite to cast them in a corporatist mould. Obviously, anti-party attitudes are not merely a result of socialist propaganda, but are deeply rooted in the country's political culture. Similarly, the Council governing Slovenian public television (RTV) is composed of only 5 parliamentary, but 17 corporate interest representatives. Here, too, there is an all but unquestioned consensus that parties are inadequate representatives of civil society and its diversity of aspirations.[23]

Since the beginning of the 1990s, an extensive system of collective bargaining has evolved. The Economic Social Council, a tripartite setup harmonizing social, labour and economic policies, concluded its first "social pact" in 1995. By now, only interest organizations recognized by the state are allowed to participate in this "social partnership". Following this logic, a huge chamber system based on compulsory membership has been set up. Some such chambers already existed prior to 1990, such as the Medical Chamber, the Chamber of Lawyers, Chamber of Social Service, and the Chamber of Nurses, which had taken over administrative competences from the state. Others had to be created anew.

Although pluralist corporatism has meant, in the past, that neither the socialist nor the Catholic pillar could exert hegemonic power, the latter has of course been greatly strengthened since Slovenia's accession to independence. Socialism (as a secular religion), Orthodox Christianity and Islam suddenly disappeared from the national scene. All of these have since been discredited by the Church as enemies of the nation. Undeniably, therefore, the Catholic church has sought to regain a position of moral, cultural and political hegemony in Slovenia. Its main political wing is the Christian Democratic Party and the new interest organisations. It has established a new daily, *Slovenec*, a grammar school in Ljubljana, a few kindergartens, and is now trying to bring confessional education back to state schools. The party scene is to some extent divided along "confessional" lines: the Christian democrats as a Catholic rural organisation opposes the Peoples Party as a non-Catholic rural organisation, the United List of Social Democrats as a strong urban and non-Catholic party has as its main foe the Social Democratic Party, which is also predominantly urban but Catholic in orientation. The strongly non-Catholic Slovenian National Party, finally, is countered by the weakly Catholic and nationalist Slovenian Right.

The Socialist pillar has been less well organized and has also largely collapsed morally. Only after a few years, when the Church displayed old political habits and renewed appetite for dominating public life, the socialist segment regained some of its former verve. It gathered all social democratic and labour oriented parties in the United List of Social Democrats (1993) and has formed strong links with the all-important Federation of Free trade Unions. But it has neither periodicals nor associations matching those of the Catholic pillar. Recently, however, it has capitalized on its association with the concept of a confessionally neutral state, education and politics.

A liberal political culture is only just emerging, and is mainly rooted in new social movements, in the movement for political and civil rights, in various civic initiatives, and in youth organizations dating back to the 1980s. Within the party spectrum, it finds expression in a faction of Liberal Democracy of Slovenia and in politically nonaligned intellectuals. On an institutionalized level, liberal elements are enshrined in the Constitution, which accords individual rights a high priority, as well as in the institution of the ombudsman, whose purpose is the safeguarding of those rights.

Conclusion

Perhaps the most pervasive and enduring pattern in Slovenian political culture has been the urge to fight the enemy to the very end. My brief historical sketch has shown that, broadly speaking, Catholicism has created, and that communism has continued this dynamic. The task of pluralist corporatism must now be to substitute that pattern by one aimed at compromise and consensus, and which includes all the relevant pillars or cultures in its system of governance. The Grand Coalition now ruling Slovenia has taken a major step in this direction by comprising the three strongest parties, which also represent the three main political orientations and segments in Slovenia: Liberal Democracy, Christian Democrats and the United List of Social Democrats.[24] At the moment, the political culture of coexistence with differences seems to be the only real possiblity of pacifying these political cultures. Parallel to this development, a lively debate on the scope for reconciling socialism and Catholicism has been led since the mid-1980s. Finally, the corporate consensus of working for the survival of Slovenia as a nation has been most evident from continuities within the political elite.[25]

Notes

1 As is well known, the concept was introduced by G. A. Almond (1956), who distinguishes between Anglo-American, pre-industrial, totalitarian, and continental-European political systems according to criteria of political culture and role structures. Unsurprisingly, the Anglo-American system reveals the most homogenous political culture as well as the most stable, differentiated, and organized structure of roles.

2 Another step towards the invention of civic culture was made by Almond and Verba (1965). They collected abundant evidence that US- and British political culture made for the best of all possible democratic polities. "Our study will suggest that there exists in Britain and the United States a pattern of political attitudes and an underlying set of social attitudes that is supportive of a stable democratic process", while other countries must be content with the possibility of "achieving a more stable and democratic polity" (Almond and Verba 1965, X).

3 Pye and Verba (1965) led the field with their study of Third World political culture compared to its Western paragon.

4 The main determinants of political culture are perceptions of time and space, and then dualities of the subjective and the objective, of the material and the spiritual, etc. These categories are often taken for granted, rather than being seen as mediated through European history as a way of strengthening a consensus governing modern society. Countries beyond the reach of modern or bourgeois time have also fallen outside the scope of today's hegemonic political culture.

5 To believe in peace and prepare for war by producing arms and training soldiers may seem like a "gap" in reasoning. Politically speaking, however, preparing for war means working for peace. Since we are not prepared to acknowledge the existing pattern of political behaviour, our only recourse is the

moral validation of war which, although nobody wants it, exists as a manifestation of evil in our bright and honest politics.

6   Gramsci (1977,1516) was absolutely right when he argued that everyone is an intellectual, while not everyone fulfills the role of an intellectual.

7   This tradition has survived in German research on political culture. See Berg-Schlosser (1987, 1993).

8   Lijphart could not resist the weight of the hegemonic conception and reformulated his classification of four types in binary terms to suit the more competitivily oriented Anglo-American blueprint (Lijphart 1984).

9   Socialist indoctrination has been discussed in these terms, but we can now see that such a procedure has been inherent in the conception of political culture since its inception.

10 Some scholars have studied political culture in schools (Židan 1993), national character (Musek 1994, Trstenjak 1991, Pécjak 1992), political actors and their values (Markič 1992, Bibič, Tomc), political culture and the formation of ideology in literature (Močnik 1981, Žižek 1987), political culture and the process of socialisation (Ule 1990), political values of youths (Miheljak,Vlado and Ule 1993) – to mention only the most important. To this we should add gender studies dealing with patriarchalism, authoritarianism, tolerance, and androcentrism (Ule 1990, Jogan 1994). Kolenc (1993) instigated a project of political culture research run by the Educational Research Institute, Ljubljana and trying to integrate the many detailed empirical studies into a longitudinal as well as cross-cultural framework.

11 The concept of politics or polity is never exclusively literal. Moreover, the metaphor is central rather than incidental to the concept. It is a powerful abstraction or myth ordering the empirical realities of politics (Lukšič 1994, 14).

12 see Zgodovina Slovencev, the History of Slovenes, for details (1979, 311).

13 Even at the peak of the Reformation, peasants largely upheld their loyalty to the Catholic Church.

14 As a consequence, Slovenian is a language extraodinarily rich in dialects.

15 The term is Ivan Cankar's, Slovenia's perhaps greatest writer, and was popularized through his novel Hlapci (servants) in which it is used as a metaphor for Slovenes.

16 According to writer Miloš Mikelen(1987, 156), exclusivism has been one of the most marked features of modern Slovenian politics. Both Catholicism and the revolutionary politics succeeding it derived their force from the determination to implement radical and thoroughgoing change. Exclusivism, on this view, is frequently found in small nations trying to create unity through erasing differences in order to protect their identity.

17 All critics of Yugoslav self-management socialism have underrated the potential for civil strife following both World War II and the year 1990. Bosnia was, and still is, among many other things, multi-party and non-socialist Yugoslavia come true. I have tried to show (Lukšič 1991) that consociative arrangements in socialist Yugoslavia tended to undergird stability and democratization.

18 The Communist Party's renaming into League of Communists in 1952 had its roots in Marx's League of Communists, but also expressed anti-party sentiments rooted especially in Slovenian political culture. Significantly, Kardelj as the main ideologist of this and other reforms of Yugoslavia's political system was himself a Slovene.

19 It is important to note that this constitution was drafted in the context of a global debate on (neo)corporatism as a possible compromise between socialism and capitalism. It was Yugoslavia's contribution to the search for a third road between (Soviet) socialism and capitalism, or between state and market regulation of society.

20 In comparison with only two years ago, however, more and more of the above associations are beginning to act like political parties.

21 Christian Socialists launched Revija in društvo 2000 (Review and Society 2000), the National Democrats set up Nova revija (New Review), Jungian Marxists and Anarchists gathered around Časopis za kritiko znanosti (Newspaper for the Critique of Science), new social movements and initiatives found a voice in the weekly Mladina, to mention only the most important groups of the 1980s.

22 For details, see especially Bibič (1995).

23 I have argued elsewhere that in Slovenia, political parties were introduced mainly because the political elite bowed to the hegemonic perception of democracy as one of competing parties.

24 These three parties hold 59 of a total of 90 seats in parliament (LDS 30, CDS 15 and ZLSD 14).

25 Janez Drnovšek was President of the Presidency of Yugoslavia and is now Slovenian Prime Minister, Milan Kučan was President of the League of Communists and is now President of Slovenia, Dagmar Šuster was President of the Economic Chamber of Yugoslavia and subsequently President of the

Economic Chamber of Slovenia until June 1995, Jožef Školč was President of the Socialist Youth Organization and is now President of National Assembly, Ivan Kristan was a member of the Constitutional Court of Yugoslavia and now serves as President of the State Council, Dušan Semolič was the secretary of the Socialist Alliance and now heads Slovenia's largest trade union, Jožica Puhar was Minister of Labour and retained that post until May 1994, to mention only the key politicians.

References

Almond, G. (1956), 'Comparative Political Systems', *Journal of Politics*, vol. 18, no. 3.

Almond, G. and Verba, S.(1965), *The Civic Culture. Political Attitudes and Democracy in Five Nations,* Little, Brown and Company, Boston and Toronto.

Almond, G.(1993), 'The Study of Political Culture', in Berg-Schlosser and Rytlewski (1993).

Berg-Schlosser, D. and Schissler, J. (eds.) (1987), 'Politische Kultur in Deutschland', *Politische Virteljahresschrift,* Special Edition.

Berg-Schlosser, D. and Rytlewski, R. (eds.) (1993), *Political Culture in Germany,* Macmillan Press, London.

Bernik, I., Malnar, B. and Toš, N. (1995), 'The Paradoxes of Instrumental Acceptance of Democracy', unpublished paper presented at the German Political Science Association' conference in Potsdam, 1994.

Bibič, A. (1995), 'Politika – še vedno umazana pesem', (Politics – Still a Dirty Song), *Teorija in praksa,* nos. 3–4.

The Constitution of the Republic of Slovenia (1992), Uradni list RS, Ljubljana.

Đordević, J. (1981), *Eseji o politici i kulturi,* Savremena administracija, Beograd.

Ferligoj, N., Rener, T. and Ule, M. (1991), 'Sex Differences in "Don't Know" Rates. The Case of Slovenia', *WISDOM,* vol. 4, nos. 1–2.

Furnivall, J.S. (1939), *Netherland's India: A Study of Plural Economy,* Cambridge University Press, Cambridge.

Gramsci, A. (1977), *Quaderni del carcere,* Einaudi, Torino.

Jalušič, V. (1992), *Dokler se ne vmešajo ženske: ženske, revolucije in ostalo* (So Far, Women are Not Involved. Women, Revolutions and the Rest). KRT, Liublijana.

Jogan, M. (1992), 'Politična kultura in seksizem' (Political Culture and Sexism), in Macura and Stanič (1992).

Jogan, M. (1994), 'Erozija androcentrizma v vsakdanji kulturi' (Erosion of the Androcentrism in Everyday Culture), *Teorija in praksa,* nos. 7–8.

Južnič, S. (1973, 1989), *Politična kultura.* (Political Culture) Obzorja, Maribor.

Kardelj, E. (1977), 'Smeri razvoja političnega sistema socialističnega samoupravljanja' (Courses of Development in the Political System of Socialist Self-management), *Komunist.*

Keber, D. (1992), 'Stroka in politika, raba in zloraba' (Profession and Politics, Use and Abuse), *Delo Sobotna priloga,* 14 August.

Kolenc, J. (1993), *Politična kultura Slovencev* (Political Culture of Slovenes), Korotan, Ljubljana.

Kuzmanič, T. (1994) 'Postocializem in toleranca ali Toleranca je toleranca tistih, ki tolerirajo – ali pa ne!' (Postsocialism and Tolerance, or Tolerance is the Tolerance of Those Who Tolerate – Or Don't!), *Časopis za kritiko znanosti,* nos. 164–165.

Lehmbruch, G. (1967), *Proporzedemokratie: Politisches System und politische Kultur in der Schweiz und Österreich*, C.B. Mohr, Tübingen.

Lijphart, A. (1968), 'Typologies of Democratic Systems', *Comparative Political Studies*, no. 1.

Lijphart, A. (1968), *The Politics of Accommodation: Pluralism and Democracy in the Netherlands*, University of California Press, Berkeley etc.

Lijphart, A. (1984), *Democracies. Patterns of Majoritarian and Consensus Government in Twenty-One Countries*, Yale University Press, New Haven and London.

Lukšič, I. (1990), 'Recepcija pojma politične kulture v Jugoslaviji' (Reception of the Term of Political Culture in Yugoslavia), *Teorija in praksa*, nos. 8–9.

Lukšič, I. (1991), 'Demokracija v pluralni družbi?' (Democracy in Plural Society?), *Znanstveno in publicistično središče*, Ljubljana.

Lukšič, I. (1992), 'Preoblečeni korporativizem na Slovenskem' (A Disguised Corporatism on Slovene Ground), *Časopis za kitiko znanosti*, no. 8, pp. 148f.

Lukšič, I. (1994a), 'Corporatism in the Political System of the Republic of Slovenia,' in Bučar, B. and Kuhnle, S. (eds.), *Small States Compared: The Politics of Norway and Slovenia*, Alma Mater, Bergen.

Lukšič, I. (1994b), '(Ne)strankarstvo na Slovenskem' ([Anti-]Partyism on the Territory of Slovenia), in Lukšič, Igor (ed.), *Stranke in strankarstvo*, SPOD, Ljubljana.

Lukšič, I. (1994c), 'Liberalizem versus korporativizem' (Liberalism vs. Corporatism), *Znanstveno in publicistično središče*, Ljubljana.

Macura, D. and Stanič, J. (eds.)(1992), *Demokracija in politična kultura* (Democracy and Political Culture), Enajsta univerza, Ljubljana.

Markič, B. (1992), 'Politične stranke, volilna kultura, volilni sistema' (Political Parties, Voting Culture and the Electoral System), *Teoria in Praksa*, vol. 29, nos. 7–8.

Miheljak, V. and Ule, M. (1993), *Mladina, druža, oklonslost* (Youth, Family, Delinquency), Center za socialno psihologijo, FDV, Liubliana.

Mikeln, M. (1987), 'Težki pokrov ekskluzivizma' (The Heavy Cover of Exclusivism) in *Razvoj slovenskega narodnega značaja v luči 4. točke programa OF*. Delavska enotnost, Ljubljana.

Močnik, R. (1981), 'Mesečevo zlato: Prežeren v ocnačevalcu', *Analecta*, DDU Univerzum, Liubljiana.

Musek, J. (1994), 'Psihološki portret Slovencev' (A Psychological Portrait of Slovenes), *Znanstveno in publicistično središče*, Ljubljana.

Novosel, P. (1969), *Politička kultura u SR Hrvatskoj* (Political Culture in Croatia), Fakultet političkih nauka, Zagreb.

Pye, L. W. and Verba, S. (eds.) (1965), *Political Culture and Political Development*, Princeton University Press, Princeton.

Rojc, E. (1975), 'Šola kot oblikovalka samoupravne politične kulture' (School as the Creator of Self-Managemental Political Culture), *Teorija in praksa*, nos. 1–2.

Reiterer, A., Halbmayer, B., Haller, B. and Lukšič, I. (1995), 'Sozialpartnerschaft in Slowenien' (Projektendbericht, Institut für Konfliktforschung), Wien.

Tomc, G. (1993), 'Slovenci o politiki in politikih' (Slovenes on Politics and Politicians), in Frane, A. (ed.), *Volitve in politika po slovenskok*, Znanstveno in publicistično središče, Ljubljana.

Toš, N. (ed.) (1995), *Dozorevanje slovenske samozavesti* (Maturation of Slovene Self-Consciousness), FDV–IDV, Ljubljana.

Trstenjak, A. (1991), *Misli o slovenskem človeku* (Thoughts on Slovene Man). Založništvo slovenske knjige, Ljubliana.

Ule, M. (1990), 'Žene i politika' (Women and Politics) Gledišta, vol. 31, nos. 1–2.

Verba, S. (1967), 'Some Dilemmas in Comparative Research', *World Politics*, vol. 20, no. 3.

Vehovar, U. (1994), 'Odsotnost moderne politične kulture na Slovenskem' (The Absence of Modern Political Culture in Slovenia), *Delo Sobotna priloga*, March 18.

Vodopivec, P. (1995), 'Glavne poteze in stalnice v slovenskem zgodovinskem razvoju' (Main Features and Constants in a Slovene Historical Development), in Rus, V. (ed.), *Slovenija po letu 1995*, FDV, Ljubljana.

Vreg, F. (1992), 'Geburtswehen einer Nation. Systemwechsel und politisch-kulturelle Trends in Slowenien', in Gerlich, P., Plasser, F. and Ulram, P. A. (ed.), *Regimewechsel: Demokratisierung und politische Kultur in Ost-Mitteleuropa*, Böhlau, Wien.

Za kulturo dialoga (For a culture of dialog), United List of Social Democrats, Conference in Velenje, 14 May 1994.

Zver, M. (1992), 'Korporativizem v slovenski politični misli v 20. in 30. letih' (Corporatism in Slovene Political Thought in Twenties and in Thirties), *Časopis za kritiko znanosti*, nos. 148–149.

Židan, A. (1993), 'Democratiňa politična kultura srednješolcev' (Democratic Political Culture of Medium-School Pupils), FDV, Ljubljana.

Zgodovina Slovencev (1979) (The History of Slovenes), Cankarjeva založba, Ljubljana.

Žižek, S. (1987), 'Jezik, Ideologija, Slovenci' (Language, Idelogy, Slovenes), Delavska enotnost, Ljubljana, pp. 9–60, and *Revija 2000*, 1989, nos. 41–42.

CROATIA

7 Croatian Political Culture in Times of Great Expectations

Pavle Novosel

Twenty-five years ago, during my early research years, I came across a book which was to influence profoundly the way I thought about politics for some time to come: Almond and Verba's *Civic Culture* (1963). Especially to a trained psychologist like myself, this book appealed since, in contrast to the then prevailing ideology, it tried to explain a number of political phenomena in psychological terms, that is, in terms of citizens' beliefs, attitudes and expectations.

Put simply, the book's main thesis is that political culture, understood as a set of beliefs and attitudes, is a major factor in the shaping of political behaviour. If people believe that they have certain rightful claims to make within a political system, they will do so, if not otherwise, by voting out the party in power. If they feel obliged to perform duties considered important for the democratic political process, they will do so irrespective of formal prescriptions and ideological proddings. In other words, Almond and Verba confirmed Luckmann's and Berger's hypothesis (1966) that people behave in accordance with their constructed social reality, and that they realize their projects in the manner of self-fulfilling prophesies.

This approach seemed to explain much of what was going on in Croatian political and social life at the time. It was, of course, the time of Yugoslavia's unprecedented experiment of "socialism with a human face": a sociopolitical system of self-management and decentralized – "self" – government based on public owenership of the means of production. This model purported to transform an authoritarian, centralized and rigidly controlled political and social system into decentralized, participatory social organizations fostering a type of democracy much akin to the participatory industrial democracy then promoted by many social theorists. For many of us, who were not far-sighted enough to envisage the possibility of communism's disintegration, this was the only hope left for a life worthy of human beings. If only for this reason, it seemed worthwhile to contribute to the development of such a socialism.

Meanwhile, self-management and self-government faced insurmountable obstacles and were not put into place as intended despite considerable pressure from the ruling party. By the late 1960s, two decades had passed since the beginning of reforms, and yet the media still carried negative reports and despondent comments demonstrating that the old autocratic arrangements were still alive and prospering, and even gaining ground in some areas of life.

Some data on Croatian political culture

For a social scientist, this was a challenge of the first order. Obviously some real and previously overlooked obstacles had to be found before any progress could be made, and Almond and Verba's study promised to be of guidance in this respect. It seemed like an obvious idea to carry out empirical research on the political culture of the country and to see how much the results agreed with a system of self-management. This could conceivably offer a more realistic basis for social planning generally, and more specifically for steps toward establishing a new kind of democracy within the socialist autocracy.

In 1968, I conducted research on a representative sample of 750 Croations, which was aimed at probing attitudes towards some critical issues of self-management (Novosel 1969). In what follows, I will present some of the results of this survey and compare them with Almond and Verba's results on five other countries – the United States, Great Britain, Germany, Italy and Mexico. Only then will we be able to put the Croation situation into perspective. Since Almond and Verba rely on a systemic theory of political processes, and since the analytical distinction between input and output is fundamental to such an approach, I will begin by presenting input data.

"Input" raises the question of whether, and to what extent, citizens believe that they are able to influence decisions taken by local and national authorities. If participatory expectations are low as a result of repeated negative experience, citizens are likely to become passive and alienated from the political system. More and more decisions will be removed from the sphere of citizen's direct influence and abandoned to the arbitrary will of bureaucrats, no matter what principles of self-government and self-management prescribe. Almond and Verba tried to gauge levels of potential civic activism by asking the question: "Suppose a regulation were being considered by a local governmental unit that you considered very unjust of harmful. What do you think you could do?" The comparative results are shown in table 1.

Table 1 Percentage of respondents unable to think of anything to do in case of harmful or unjust local government action

Nation	percentage of negative answers
Americans	23
Britons	22
West-Germans	38
Italians	49
Mexicans	47
Croats	47

Note: this and all following tables are compilations of Almond and Verba's and my own results.

The data show that only Italians displayed a more negative attitudes than Croats, the latter being on a level with Mexicans. Almond and Verba, of course, chose Mexico as an example for unjust and illegitimate governance, in which an all-powerful "institutional-revolutionary party" had been monopolizing the political process for many years. From a Croation perspective, these results clearly pointed to the correspondence between negative political experience and despondent attitudes with respect to political participation. They also highlighted that such accumulated experience was likely to become a serious obstacle to Croation democratic self-management, since Croats' experience with participation within the Yugoslav federation had been mostly negative.

A similar question aimed at probing the degree of alientation from decisionmaking at state levels: "Suppose a law were being considered by a federal legislature that you considered to be unjust or harmful. What do you think you could do?

Table 2 Percentage of individuals who could not think of anything to do in case the federal legislature were considering an unjust or harmful law

Nation	percentage of negative answers
Americans	26
Britons	38
West-Germans	63
Italians	72
Mexicans	62
Croats	73

Thus table 2 shows that with respect to federal institutions, Croats top the list of people with no hope of influencing government decisions. In part, this may be explained by the fact that Croats had been second class citizens in the context of the Austro-Hungarian monarchy, within the Kingdom of Yugoslavia (1918 to 1941), and in socialist Yugoslavia. In the latter especially, they were victims of a twofold dynamics of exclusion: the chasm between communist elites and the rest of the population, and stigmatization as the "defeated fascist nation".

Questions relating to the output side of a political system yielded, predictably, similar results. To the question "Suppose there were some question that you had to take to a government office – for example, a tax question or a housing regulation. Do you think you would be given equal treatment – I mean, would you be treated as well as anyone else?" the answers were as follows:

Table 3 Percentage of respondents not expecting to be treated equally by administrative bodies

Nation	percentage of negative answers
Americans	39
Britons	13
West-Germans	28
Italians	30
Mexicans	55
Croats	63

Croat experience and corresponding political attitudes in this respect were by far the worst. Discrimination by administrative bodies seemed to them, for the most part, an established and incontestable fact, and this had inevitable repercussions on their political culture. There is more to come, however. Croats knew very well where to look for the sources of their plight, blaming either the republic's government at Zagreb or the federal government in Belgrade. This could again be shown clearly by means of the following question, which aimed at eliciting views on the efficacy, or systemic output, of government: "On the whole, do the activities of the national goverment tend to improve conditions in this country or would we be better off without them?"

Table 4 Percentage of respondents with negative views on national government efficacy

Nation	percentage of negative answers
Americans	22
Britons	18
West-Germans	33
Italians	25
Mexicans	37
Croats	54

Here, the gap between Croats and citizens of other nations sampled by Almond and Verba is even more marked, with even Mexicans scoring a remarkable 17% lower.

Given these results, it is hardly suprising that the movement for national independence gained such momentum and included a substantial portion of the ruling communist elite in Zagreb. It was widespread disillusionment with government which fuelled both the Croat Spring of the late 1960s and the first free multi-party elections in

1990, brought about by "young lions" within the Communist Party that was subsequently removed from power.

From such data on experienced and anticipated discrimination, it would be easy to jump to the conclusion that Croats harbour resentments against other nations of the former Yugoslavia, and especially against their own big brother toward the East. But surprisingly, this is not the case. 81% of Croats insisted that they would help someone in need irrespective of his or her nationality, and only a minority of 16% openly declared that they would restrict such assistance to a member of their own community. Catholic morals in a predominantly Catholic country might go some way towards explaining this, and it may in any case be worth trying to improve the image of a peaceful nation largely tainted by the role it played during World War II.

Great expectations and their fulfillment

Now that an independent state with a freely elected government has finally been achieved, the question of Croat political culture's further development naturally arises. It is a question now being asked throughout Eastern Europe's former communist states: will the new political context result in the emergence of a civic culture – according to Almond and Verba an indispensable prerequisite for maintaining a workable democracy – or will alienation from the political system and despondent attitudes with respect to participation persist?

Judging from the first five years of the new state, neither possibility can be ruled out. Since up to date research from which to draw conclusions is relatively scarce (cf. Šiber 1992 and 1993, Vukadinović 1992, Zakošek 1994, Tomac 1944, Leinert-Novosel 1995, Ogorec 1994, Tadić 1944, Kasapović 1993, Županov 1995) the following considerations are of necessity subjective and unsubstantiated. Some facts, however, are simply too obvious to be denied, and so, too, are perhaps some of the arguments in my ambiguous prognosis. I think we can expect an initial period during which things will remain the same, and may even get worse. But once the changes in the political system have produced changes in the way people experience politics and its effects, political culture should begin to adapt to altered circumstances. There is, in principle, no reason why Croatian democracy should not slowly develop along the lines of at least some traditional democracies.

This prediction is based on a comparison of today's Croatia with that of 1990, when the old regime was replaced. The changes were originally almost universally welcome. To a psychologist, the mood of optimism, the hailing of freedom and independence and the castigation of socialist as well as other, previous, wrongs seemed very much like a bout of collective euphoria. The future seemed bright to just about everyone, great expectations would at long last be fulfilled.

Exhilaration of this kind is known to have occurred, for instance, during the French Revolution, in the course of the establishment of some Eastern European socialist regimes just after World War II, or in some African and other Third World states at the moment of gaining independence from colonial rule. It is an intense but transient feeling of delivery at a moment during which individual and collective antagonisms seem to be resolved once and for all. Naturally, such naive beliefs must soon clash with harsh realities, in Croatia as everywhere else. In the case of Croatia, that clash perhaps came sooner and more forcefully in the face of many adverse circumstances.

To mention only a few of them, about one third of the country was under military occupation and could not be incorporated into the new state. More than 700 000 refugees had left their homes. Even the cardinal question of national independence, therefore, defied resolution. Living standards fell to roughly one fourth of what they had been before the war. In addition, privatization of the economy exposed the ugly side of capitalism: opportunists were acquiring "public" companies for ridiculously low sums, which distinctily smacked of corruption and downright ransacking of public property. Some began to ask themselves where exactly the difference between old and new regimes lay, and could not help seeing the continuity of a chasm between citizens with privileges and those without them.

Social security deteriorated as overstaffed plants layed off their workers and unemployment rose to about 17 percent of the workforce. Some of the new entrepreneurs began to behave in ways reminiscent of times prior to the emergence of trade unions. Pensions fell to unprecedentedly low levels, and many elderly people survived only with Caritas assistance after a working life of 30 or 35 years.

Lack of experience in the new administration and political elite did nothing to alleviate these grievances, and suggest that deficiencies of political culture are at least as rooted among the governing as they are among the governed. For instance, some opposition parties boycotted Parliament after electoral defeat in 1990 and 1992 on the grounds that they were not granted the means of influencing government policies. Having been outvoted on several occasions, they simply walked out proclaiming that the elected Parliament would henceforth be illegitimate.

The ruling Croation Democratic Union (Hrvatska demokratska zajednica) recently demonstrated its poor understanding of what communication between a government and an attentive, participatory citizenship entails. While traditional democracies have long recognized the need to keep open channels of communication as a means of gaining public support, the Croation government has demonstrated with alarming frequency its inability to understand this.

By way of example, the government announced a change in health care regulation restricting the number of prescriptions issued to each patient. Since unlimited coverage of prespription was becoming too expensive and was also widely abused by physicians and patients alike, this was a more then sensible proposal. But the new rules were proclaimed in an incomprehensible bureaucratic jargon causing people to panick: for weeks, patients virtually laid siege to practices in order to obtain prescriptions they would probably never need. Only under intense pressure from the media, the Minister of health finally explained that no one depending on medication would have cause to worry. Whatever the intricacies of this question, it is clear that the poor flow of communication has been dealing many blows to the fragile legitimacy of the new political system. For the time being, we can only guess at the origins of this phenomenon. One reason is perhaps that bureaucrats are so overworked that they have no time to explain what they are doing. Another, more convincing explanation could be that the notion of public accountability has never been rooted in Croat political culture, and that the new political representatives are unable to handle or deal with the limelights of civic attention. Be that as it may, leaders are quickly exhausting their credit of public confidence by acting in the way they have done on many occasions. In many instances, citizen's system affect is undermined rather than fostered, and with it the development of democratic political culture.

Conclusion

As in other former socialist countries, the birth pangs of Croatia's new democracy have considerably thwarted the emergence of a new political culture. Initial expectations were too high for any new government to fulfill, and the resulting disillusionment poses a threat to democratic consolidation. It potentially reinforces a political culture of alienation inherited from the past; and it may even contribute to a partial reinstatement of authoritarian rule.

Part of the answer to such problems might be found in improved communication. Explaining a situation candidly and coherently, defending openly the inevitability of new rules and laws, countering suspicions and despondency, and, perhaps most importantly, acknowledging that democracy is only the best among several flawed options, could help promote realism and channel expectations into activism. The immediate future should ideally be a period of increasing sobriety rather than disillusionment, a period of gradual acceptance of democracy's imperfections coupled with the will to improve them as much as the situation allows. If, in contrast, the new democracies abandon the field to a politics of collective spontaneity, something quite different to democratic political culture will develop in the new free states of Europe, Croatia included.

References

Almond, G. and Verba, S. (1963), *The Civic Culture. Political Attitudes and Democracy in Five Nations*, Princeton University Press, Princeton.

Berger, P. and Luckmann, T. (1966), *The Social Construction of Reality*, Doubleday, New York.

Kasapović, M. (1993), *Izburni i stranacki sustav republike hrvatske* , Alinea, Zagreb.

Leinert-Novosel, S. (1995), 'Političko stranke i žene', *Politička misao*, no. 2, vol. 32, pp. 112–140.

Novosel, P. (1969), 'Politička kultura u SR Hrvatskoj' (Political culture in the Socialist Republic of Croatia), *SSRNH*, pp. 64ff.

Ogorec, M. (1994), 'Hrvatski domovinski rat', *Opatna Otokar Hrsovani*.

Šiber, I. (1992), 'Politička kultura i tranzicija', *Politička misao*, no. 3, vol. 29, pp. 93–110.

Šiber, I. (1993), 'Structuring the Croatian Party Scene', *Politička* misao, no. 2, vol. 30.

Tadiæ, V. (1944), *900 Dana Rata – U meduvjerskim odnosima*, Birotisak, Zagreb.

Tomac, Z. (1944), *Tko je ubio Bosnu: Iza zatvorenih vrata (drugi dio)*, Birotisak, Zagreb.

Vukadinović, R. (1992), *The Break Up of Yugoslavia: Threats and Challenges*, La Haye, Clingdale.

Zakošek, N. (1994), 'Struktura i dinamika hrvatskog političko sustava', *Revija za sciologiju*, nos. 1–2, vol. 25.

Županov, J. (1995), 'Masovni mediji i kolektivno nasilje, *Politička misao*, no. 2. vol. 32.

8 The Sleeping Beast. Central European and Croatian Political Culture

Josip Kregar

> Water freezes, ice thaws, iron rusts, iron oxide decomposes; hay becomes beef, beef may become hay again; revolt and reaction are cyclical and opposite processes in society (...) But the temporal order of events remains immutable; it cannot be reversed. The historic process and evolutionary process are alike in being temporal in character, i.e. non repetitive and irreversible (...) From the point of view of the evolutionary process every historical event is an accident and in this sense unpredictable. (White 1949, 13–14)

Traditions and perspectives

As part of an ancient Roman ceremony, war heroes were allowed to display their spoils, so as to impress all with the power of Rome. Throughout this public display, however, captives now reduced to live booty or slaves constantly repeated a single chant: *Te hominem esse memento* – remember that you are only human. Throughout history such voices have been subdued, and few have cared to remember their humanness in times of triumph. We are now at just such a time of triumph: the victory of democracy over totalitarianism, of interdependence over isolation, of inclusion over exclusion, and of integration over disintegration; a time at last for freedom and choice. We are, it is said, at the end of history (Fukuyama 1992).

However, the abundant spoils have not been equally distributed across Europe. Progress in some parts of the continent fall short of expectations and instead of a bright future, what we see is a decline of morale, economies in shambles, social tensions, and exacerbated political conflict. Not least, we ought to remember that in many parts of Eastern Europe, people literally struggle to survive.

While we may say that history is beginning anew in the East, we must recognize that traditions are still weighing down heavily on us. We are hemmed in by old enemies: undemocratic political traditions and culture; destruction of the moral foundations of society; corrupt and savage political elites; and inadequate as well as inefficient democratic institutions.

Political culture in post-socialism

Central Europe's political culture is the fruit of war memories, of servitude to lords, kings and decaying empires, of euphoric liberations and victories, of bureaucratic routine and the apathy it has engendered. But despite such seemingly overriding similarities, Central Europe is far from being culturally or politically homogeneous. It is, on the contrary, a maze of different nations, minorities, cultural patterns, cuisines, religions, languages.

The political culture of post-socialist societies differs from Western patterns largely as a result of forty years of Leninist "democracy". Although a number of arguments support the assumption that such divergences will disappear in the future, they are, at least for now, an incontestable reality. As one commentator has put it, "[p]erhaps the beginning of wisdom is to recognize that what communism has left behind is an extraordinary mish-mash." (Ash 1992, 286)

Central Europe's present chaos of political values is, for the most part, a product of communism's monopoly of power and the means of ideological indoctrination. Not only was the "dictatorship of the people" a very real one, and one extremely brutal at various times and in various places; ideological penetration could also reach deep into everyday life, and it pervaded political institutions as well as public opinion. One's career, and sometimes one's life, depended on utter hypocrisy (Volensky 1984) and erosion of moral standpoints: at any moment, it was necessary to distinguish between the public "virtue" of ideological obedience and opinions held privately.[1] Clearly, living under such conditions could not fail to have severe social and psychological repercussions.[2]

Of course, the specificities of Central European political culture cannot be explained by reference to post-war developments alone. Dinko Tomasic saw this clearly. He suggested a distinction between "Dinaric" and "Pannonian" forms of life in an attempt to take account of long term cultural development in the region (Tomasic 1948). The former he described as a tradition of militant authoritarianism, favouring values of authority, leadership, religion, and nation.[3] Individual life, in this pattern, is of secondary importance, while risk-taking, sacrificial virtues, and idealism, however rhetorical, are instilled as primary goals during many stages of socialization. Notions of courage, petulance, passion and "dionysiac" euphoria, of heroic effort, honour and *thymos* all fit into this category.

> Thus we find that most states in Eastern Europe, from medieval times up to the present day, were and are ruled by despotic and oligarchic governments. (...) After World War II, when most of the countries in Eastern Europe were occupied by the Red Army, the autocratic and tyrannical practices inherited from the past, and supported by the self-maximizing tendencies of the mountain folk and the local military and intelligentsia, were only accentuated. In some cases, the practices were legalized in the form of new laws and constitutions. (Tomasic 1948, 117, 126)

Pannonian culture, in Tomasic's terminology (1948, 228), describes a more "Apollonian" social order of artisans and farmers, whose predominant values are rationality, hard work, diligence, respect for material goods, and security within the family or village community. "In contrast to the insecure economic and social conditions of stock breeders in the mountains, the plowmen who settled the lowlands

116

enjoyed an autarkic economy, an equalitarian family system, and remarkable lack of sharp social differentiation. Since there is more personal and economic security and more social equality in this society, there is less incentive to struggle for status, less social mobility, and less instability." Pannonic civilization promotes obedience to fathers, priests, kings and states. Cooperation for mutual benefit and peaceful compromise are valued highly in this cultural pattern, and isolation from larger frameworks of the state, nation, and region are seen as beneficial.

It would of course be an undue simplification to conclude, for instance, that the Dinaric pattern adequately describes Russians, Hungarians, Croats, Serbs, Montenegrins, or Baltic nations, or that Panonian culture denotes Bavarians, Austrians, Czechs, Slovaks, Slovenes and Poles. If cultural patterns can at all be ascribed to nations in this way, then only in order to highlight the persistence of dual and seemingly contradictory traditions of obedience and insubordination, of withdrawal into privacy and public activism. Central European political culture and its main problems can perhaps be said to have emerged from the interstices or frictions between Dinaric and Panonic traditions, which have produced, on a purely political level, a relatively consistent pattern of mild conservatism and orientation toward traditional values.

Ultimately, the mish-mash has resulted in something rather different to the sort of civic culture that seems necessary for generating the dynamics of liberal democracy.[4] The blueprint of "polyarchy" (Dahl 1956) and its dynamism, as well as that of active and discursive competition between two parties, seems at present like an anomaly in a Central European context.

Some features of Croation political culture

Empirical research conducted between 1976 and 1989 in Croatia has identified five basic types of value orientation (Petkovic et al. 1990; Ivanisevic et al. 1986).

- Traditionalist (21.63%) with positive attitudes towards religion and customs, negative views on modernization, urban life, consumerism, and equality between the sexes.
- New Right (37.40%) favouring capitalism, a liberal market economy, entrepreneurship, and self-interest, while rejecting socialism and non-economic individualism.
- Old Left (15.16%), which may be described as a Stalinist mindset.
- Innovative (23.13%) looking toward technological improvement as a source of progress. Neutral with respect to political questions, Negative attitude towards tradition, monopoly of political power, and charismatic leadership.
- Nonconformist (2.68%) exalting individual choice of life styles while rejecting tradition, religion, and conservative order.

More recent research (Petkovic et al. 1990, 577–618; Kregar et al. 1990, 619–637; Radin et al. 1990, 99–106) conducted on the eve of Croatia's democratization has, however, produced slightly different results[5]. When the distribution of values was computed around the axis of "conservative" and "innovative" and the more narrowly political polarization of "left" and "right", Croations were found to have predominantly conservative values while declaring themselves equidistant to "right" or "left".[6] Of particular interest is the fact that there seems to be a strong connection between conservativism and leftist orientations, as well as between conservatism and authoritarianism.[7]

It must be said that, on the basis of these results, it remains questionable whether Croatia's new constitution provides sufficient foundations for constructing Western-style democratic institutions within the near future. Preference for conservatism, authoritarian leadership and a rigid social order hardly provide a firm ground for democratization. But there is still cause for long term optimism. If liberal democracy and individualism are consolidated globally, there is no reason why post-socialist societies should not follow a similar trajectory of political development.

Destruction of the society's moral foundations

One of the most disruptive features of post-socialism has been its lack of shared value orientations. Social interaction is emptied out by a pervasive relativism. There is, quite simply, nothing to believe in. Socialist dogmas have been destroyed by their own incredibility, but these same dogmas succeeded in eliminating traditional moral foundations of society. Cynicism, doublethink, and an erosion of moral norms have spread under the influence of values imported from developed societies. It is not so much that a "Western way of life" has caused disorientation by penetrating our societies, but that it has seeped through in false images of pure leisure and entertainment, of welfare for all, of a society in which everyone is young, healthy, and busy nurturing his or her emotions, of a society full of animation and excitement. Such flamboyant images have, predictably, outshone those of technological superiority, of hard work, and of knowledge.

This caricature of a "Western way of life" has ended up hypnotizing the poor and disoriented, and has caused symptoms of anomy: material aspirations cannot be fulfilled by regular means, but only through deviant behaviour (Merton) and/or in the course of a general collapse of social norms (Durkheim)[8]. Obvious reactions to such a social enviroment have been the search for new orientations, and specifically the urge to unite and identify with new emerging leaders, cliques, clans, and nations. Many people, to be sure, have found a way back into a self-defined normality, but disorientation and a sense of insecurity undoubtedly remain endemic phenomena in post-socialist societies.

Institutions

Jurisdiction and government – as crucial regulatory mechanisms – have been burdened with expectations that are simply impossible to meet, at least within the short term range. Radical transformation of the entire legal system has been accompanied by technical imperfections, uncritical application of institutional solutions from abroad, and numerous legal antinomies. As a factor of social stability, the government is often countervailed or thwarted by the inefficiency and size of its bureaucracy[9], as well as by the widespread distrust which government officials encounter as exponents of the old regime.

The new institutions – parliament, political parties, administration – in no way correspond to the paradigms these terms invoke. Parliament, no matter how strong the concentration of power, should prevent its monopolizing as well as mistakes made through haste. The President, no matter how authoritarian and vile (or wise), is elected by the people and should be accountable to the people. The judiciary should strive to be independent, and should submit to the law rather than to political whim. Instead, tendencies towards presidentialism, oligarchization within political parties, circumvention of Parliament and public debate in decisionmaking, and abrupt as well as inscrutable changes in the institutional framework are characteristic of the present situation in Croatia.

But these shortcomings in democratic practice are not immediately obvious. The constitution and legal structure are comparable or even indistinguishable from those in consolidated democracies. Institutionalized civil rights and elections, a constitutional court and parliament, a formally independent judiciary and supposedly transparent administration all suggest an accomplished transformation to a modern democracy. The reality, however, looks somewhat different. Precise rules and their consistent interpretation, and generally equal treatment of citizens before the law are the exception. As a rule, rules are simple décor. They are often far too complex to be put into practice, and there is neither a routine of applying them nor a habit of being subjected to them.

It has to be said that, without pragmatic "reinterpretation" of procedural rules, very often nothing would work at all. A degree of flexibility in the application of new rules derived from high-flying declarations is an unavoidable tribute paid to the reality of a gradual transformation. While the system may well emulate Western paradigms, the pace of change will vary considerably with place and time – and will come to be recognized as a major variable in transformation processes. But divergences in development should be seen as grounded in differences of culture and tradition, or of dynamics and constraints, rather than as differences in general course and direction.[10] According to Franklin and Baun

> constitutionalism or, more simply put, the rule of law, is more likely to emerge in a state from a fertile, organic environment or political culture (...). Any institutional design will be inadequate for the development of self-sustaining democratic institutions. This proposition rest on a very special theory of constitutionalism; i.e. that a constitution is much more than a document which lays out a set of laws for the design of government. Most states, including dictatorial ones, have some form of written constitution, and it is not the written constitution in these states that has failed. (Franklin and Baun 1995, 2)

Legitimation in post-socialist societies is often pre-democratic, and neither tradition nor political culture support peaceful and incremental development. These societies are divided in a variety of ways, and people are keen to improve their personal situation and material living standards quickly. In many ways, the constitutional and legal system does not correspond to informal codes of behaviour.[11] Governments come and go, constantly trying to change a desperate situation through acts of legislation. But the problem is not primarily one of imperfect regulation or lack of administrative expertise. Rather, the conditions for reception and acceptance of these rules are unfavourable as a result of the fact that political culture and political system no longer match.[12] The gap will only be closed, however, if institutions and historical cultural patterns begin to be mutually supportive of, and adapted to, each other. Such a time will surely come, but it is just as clear that we have a long way to go before it does.

Before they can function, institutions need at least a minimum of durability in order to develop a precise understanding of their competences and objectives. In the short run, human resources are perhaps the most pressing problem. Senior officials discredited as henchmen of the communist regime have been replaced by newcomers often educated but inexperienced, or revolutionary in intent but lacking in specialization. More troubling still, the work ethos and other values of the new personnel reflect the intractable inertia of the old system. Everything somehow seems located "in between"; a reformed inertia is confronted with ideas and inspirations from abroad; new knowledge and expertise is contrasted with politics as usual; while change is expected, stability proves at times necessary, and at times overwhelming (Kregar 1964).

A new class on the scene?

Change is often interpreted as the result of clashes amidst rivalling political elites (Djilas 1957). The old Central European nomenclature lost its struggle against ambitious and vigorous dissidents, who were supported by people disappointed with socialism and its economic disaster, by moral degeneration and political inefficiency.[13] But the change has been from one elite to another. Biographies of the new leaders, the structure of the new parliaments, trials against prominent former communist politicians, and the emerging *nouvelle richesse* show clearly that the main obstacle to faster change is an inadequate selection of the political elite. We have either ended up with "tacit communists" in the new institutions, or have had to face problems arising from a new elite trying to impose its own order of priorities. These difficulties have complicated denationalization and privatization, and, to some degree, help explain the chronic shortage of personnel and competent managers needed to run the new economy. They have also hampered national emancipation and identification.

A genuinely new direction of development will only come about through all-out privatization of the economy, fundamental changes in legislation, imitation of Western-generated lifestyle patterns and radical reforms in education. So far, not much has changed, and we are instead confronted with a new ruling elite seeking legitimation through alternative ideologies such as nationalism, imperialism, and/or other civic religions.

> It may happen in the history of a nation that commerce with foreign peoples, forced emigrations, discoveries, wars, create new poverty and new wealth, disseminate knowledge of things that were previously unknown or cause infiltrations of new moral, intellectual and religious currents (...) Once [this] has set in, it cannot be stopped immediately. The example of individuals who have started from nowhere and reached prominent positions fires new ambitions, new greed, new energy and this molecular rejuvenation of the ruling class continues vigorously until a long period of social stability slows it down again. (Mosca 1961, 602)

The new class structure reflects new interests hiding behind the facade of new institutions. Post-socialist societies are not divided along the lines of political opinions and visions, but along those of wealth, power, and status. This emerging polarization of society needs to be justified, but neither traditional values nor inhererited political culture can provide a basis for such legitimation. The new system does not yet have sufficient wealth to distribute, or political power to abuse. Entrepreneurship is, at present, a skill of capitalizing on legal loopholes or of importing luxury commodities, rather than an attempt at careful calculation and production for a world market. Intellectuals, journalists, or the Catholic clergy were relatively influential five years ago, but have again been marginalized. At least partly as a result of all these failures, civil society is not emerging in ways typical of liberal democracies.

However far-fetched, a comparison with post-Napoleonic Europe seems irresistible. New states and societies are emerging on the edge of Western Europe. These states are rich in inhabitants, territory, and natural resources, they are new and unsaturated markets, and they are as powerful as they are unpredictable. Two hundred years ago, Western European governments faced with a comparable situation were prudent enough to try and achieve a new balance between principles of nationhood and constitutionalism, of public order and accountability (Allot 1993, 177-211). But how are we to find a balance now? How can nationalism be accomodated? Where should we look for strong foundations of legitimacy? By what means can we guarantee public order? And, finally, how can we devise laws that will take hold in the minds and hearts of the people?

Perspectives

Although the political and social changes at the end of 1980s were motivated economically, cultural, social and psychological causes were equally important.

> The fundamental impulse for the reforms undertaken in the Soviet Union and China was indeed economic. It lay in the inability of centralized command economies to meet the requirements of 'postindustrial' society. But even when we accept this as a long term explanation for the breakdown of communism, we cannot understand the totality of the revolutionary phenomenon unless we appreciate the demand for recognition which accompanied the economic crisis. People did not go into the streets of Leipzig, Prague, Timisoara, Bejing, or Moscow demanding that the government give them a postindustrial economy. Their passionate anger was aroused over their perceptions of injustice, which had nothing to do with economics. (Fukuyama 1995 25)

Individual and collective passions are channeled and restrained by a complex interaction of cultural values, ethical norms, and laws. As has been shown, it is precisely this process which has been more than temporarily disrupted, and whose relaunch needs to be recognized as a political task. More often than not, sombre prognoses are merely a projection of temporary obstacles or personal prejudices onto the future. But the same is true of optimistic predictions, and if we take into account all presently identifiable material constraints (economic conditions, political development, foreign interventions) and nonmaterial factors (traditions, social memory, individual aspirations), there are simply too many contingencies for plotting the course of future development.

As a result of both the tremendous backlog and the lack of experience in problem solving, new remedies for the ills of the new nations are being devised all the time[14]. Most of these solutions are impromptu reactions and do not provide anything approaching a plausible, linear, long term perspective. It should be obvious by now that we are far from the end of history, and that the history of post-socialist pre-democracies has only just begun. Manifestly, these are not times of triumph. Our foes of authoritarian political traditions, social anomy, corrupt and savage political elites, and inefficient institutions may be dormant now, but they may rise one day as horrid ghosts of the past.[15] It will take bravery in every sense of the word to face them. "Men [and women] of real fortitude, integrity and ability are well placed in every scene; (...) they show that while they are destined to live, the states they compose are likewise doomed by the fates to survive and to prosper" (Ferguson 1966, 280).

Notes

1   This should not be generalized to the dictum that "tyranny is produced obedience" (Montesquieu). Selection and behaviour of elites changed over time. The "dual executive type of career" (Fisher 1968) and modernization underwent profound changes towards the "end of ideology".

2   "Through their organization and ethos [Leninist regimes] have stimulated a series of informal adaptive social responses (behavioral and attitudinal) that are in many respects consistent with and supportive of certain basic elements of the traditional political culture in these societies." (Jowitt 1992, 287)

3   "Strong drives toward selfmaximation, reliance on violence and craftiness, and extreme hostility and factionalism were the outstanding traits of the warriors of Ural-Altaic origin" Tomasic, 1948, 106).

4   "Driven by the powerful urge toward self-maximation, these warriors were always ready to take advantage of the rivalries of the great powers by gaining and expand local political control. Once in position of power and control, they used first religious and later nationalist ideologies to unify and assimilate their subjects and to consolidate their rule." (Tomasic 1948, 115)

5. These were also differences in statistical analysis. We began with factor analysis, based on 24 factors later reduced to the following: left ideology, technocracy, youth values, religion, new social movements. Proceeding at first inductively, we then defined four different clusters on a continuum of conservativism/innovation (conservativism, neutral, innovative, radical), and political orientations (left, neutral, right). We also measured the psychological profile of respondents along criteria of authority, rigidity, machiavellism, intensity of identification, anomy, etc. Multivariant analysis yielded the results presented here.

6   39,4% of respondents are predominantly conservative, as against 27.8% who are innovative. 19.9% are neutral with respect to social and political change, while 12.8% favour it in radical form. In terms of political allegiance, 31.2% are left, 31.5% right oriented while a majority remained neutral. It is of course possible that value orientations do not logically correspond to political orientation. Thus leftists are conservative, rightist innovative, and both more or less rigidly authoritarian.

7   Very high was also the correlation between defined clusters of attitudes and nationality, education and profession. Roughly, higher education corresponds to greater openness to innovation and is not related to any one political orientation. Peasants and workers are conservative and right oriented, an average of Serbs are leftist and conservative.

8   "In the case of economic disasters, indeed, something like a declassification occurs which suddenly casts certain individuals into a lower state than their previous one. Then they must reduce their requirements, restrain their needs, learn greater self-control (...). So long as the social forces thus freed have not regained equilibrium, their respective values are unknown and so all regulation is lacking for a time. The limits are unknown between the possible and the impossible, between what is just and what is unjust, between legitimate claims and hopes and those which are immoderate. Consequently there is no restraint upon aspirations (...) Appetites, not being controlled by public opinion, become disoriented, no longer recognize the limits proper to them (...) The state of deregulation or anomy is thus further hightened by passions being less disciplined precisely where they need more disciplining." (Durkheim 1961, 920–921)

9   "The Balkan official regards himself as immeasurably superior to the peasants, among whom he lives and from whose ranks he has sprung. To be an official is the fondest dream of every able young son of a peasant. The balkan official does not like to work." This is not only a characteristic Balkan, but a general East European feature. (Polonsky 1975, 6)

10 "The comparative method means primarily intercultural comparison" (Waldo 1969, 18).

11 "Legal rules are treated as obstructions to be by-passed informally. The standard explanation for the variance is that the enforcers must make an informal exception because they do not have (and do not

122

want) the information upon which to make a rational decision. The ambiguous quality of the rules is compounded by the extraordinary mixing of traditional myths with rational standards (...) Since there is little broad-scale agreement upon the basic norms of society and many groups remain unassimilated into the nation, it is terribly difficult to get everyone to agree to abide by standard legal formulas. Control must be grabbed – through coercion, violence, money, or charismatic rule, but rarely through constitutional authority." (McCurdy, 1977, 322)

12 "The people subject to regulation become indifferent to prevalence of non-conformity with policy. Policy makers, exasperated with an intractable situation, try to correct it by drawing up more rules and passing more laws, which remain as formalistic as their predecessors." (Riggs 1964, 17).

13 "Revolutions come about through accumulation in the higher strata of society – either because of a slowing down in class-circulation, or from other causes - of decadent elements no longer possessing the residues suitable for keeping them in power, and shrinking from the use of force; meanwhile in the lower strata of society, elements of superior quality are coming to be fore, possessing residues suitable for exercising the functions of government and willing enough to use force" (Pareto 1961, 555).

14 Sometimes old ideas reappear in a new garb. "From the point of view of the interests of the peoples in this part of Europe, and from the point of view of peace in the world, it seems that the solution of the problem of Eastern Europe can be found only in internationalization and the elimination of the dominance of the military in these countries (...) But as a precondition of successful internationalization and demilitarization, the Balkan countries must advance economically and become politically independent from the great powers. (...) Demilitarized, democratized, economically and politically independent, culturally integrated, and educationally advanced, the states of Eastern Europe could be grouped together" (Tomasic, 1948, 237).

15 I am not referring specifically to the former Yugoslavia.

References

Allot, P. (1993), 'Self Determination-Absolute Right or Social Poetry in Modern Law of Self-Determination', in Tomuschat, C., *Modern Law of Self-Determination*, Martinus Nijhoff Pub, Dordrecht.

Dahl, R.(1956), *A Preface to Democratic Theory*, University of Chicago Press, Chicago.

Djilas, M. (1957), *The New Class: an Analysis of the Communist System*, Praeger, New York.

Durkheim, E. (1961), 'Anomic suicide', in Parsons, T. et al., *Theories of Society*, Free Press, New York, pp. 920–921.

Ferguson, A. (1966), *An Essay on the History of Civil Society*, ed. Duncan Forbes. Edinburgh.

Fisher, G. (1968), *The Soviet System and Modern Society*, Atherton Press, New York.

Franklin, D.P. and Baun, M.J. (1995), *Political Culture and Constitutionalism*, Sharpe, New York.

Fukuyama, F. (1992), *The End of History and the Last Man*, Free Press, Toronto.

Fukuyama, F. (1995), 'On The Possibility of Writing A Universal History', in Melzer, A.M., Weinberger, J., Zinman, R., *History and the Idea of Progress*, Cornell University Press, Ithaca, pp. 13–30.

Ivanisevic, S. et al. (1986), *Uprava i drustvo* (Administration and Society), IDIS Zagreb.

Jowitt, K. (1992), *New World Disorder*, University of California Press, Berkeley.

Kregar, J. (1994), 'Corruption in Postsocialist Countries', in Tang, D.V. (ed), *Corruption & Democracy*, Institute for Constitutional & Legislative Policy, Budapest.

Kregar, J. et al. (1990), 'Mjerenje rigidnosti: testiranje instrumentarija faktorskom analizom na poznatim grupama' (Measurement of Rigidity: Testing Selected Groups) *Zbornik Pravnog fakulteta u Zagrebu*, no. 40., pp. 619–637.

McCurdy, H. (1977), *Public Administration: A Synthesis*, Cummings, Toronto.

Mosca, G. (1961), 'On the Ruling Class', in Parsons; T. et al., *Theories of Society*, Free Press, New York, pp. 600–608.

Pareto, V. (1961), 'The Circulation of Elites', in Parsons 1961, 550–563.

Polonsky, A. (1975), *The little dictators: The history of Eastern Europe since 1918*, Routledge & Kegan Paul, London.

Petkovic, S. (1977) *Vrijednosne orijentacije prema drustvenim promjenama* (Value Orientations towards Social Change), IDIS, Zagreb.

Petkovic, S. et al.(1990), 'Konzervativnost i ideologije u Hrvatskoj krajem 80–tih godina' (Conservativism and Ideology in Croatia at the end of the 1980s), *Zbornik Pravnog fakulteta u Zagrebu*, no. 40, pp. 577–618.

Radin, F. et al. (1990), 'Neke mjerne karakteristike ljestvice autoritarnosti' (How to Measure the Authoritarian Personality), *Primjenjena psihologija*, 11, 2, 1990, pp. 99–106.

Riggs, F. (1964), *Administration in Developing Countries*, Houghton Mifflin, Boston.

Tihany, L. (1976), *A history of Central Europe*, Rutgers, New Brunswick.

Tomasic, D. (1948), *Personality and Culture in Eastern European Politics*, Stewart, New York.

Volensky, M. (1984) *Nomenklatura*, Doubleday, New York.

Waldo, D. (1969), 'The Theory of Organization: Status and Problems', in Etzioni, A., *Readings in Modern Organization*, Prentice-Hall, Eaglewood Cliffs.

White, L. (1949) *The Science of Culture*, Grove Press, New York.

HUNGARY

9 Political Culture and System Change in Hungary

Attila Ágh

Institutional versus cultural approaches in system change

The most marked paradox in Hungary's (or, for that matter, Central Europe's) democratic transition has been the contrast between rapid and profound institutional changes on the one hand, and the continuity of a stubbornly static political culture on the other. By shaping the institutions of the new political system, political culture – a set of competitive subcultures – has dominated system change from the beginning. Political institutions, which are normally seen as determinants of any *de facto* political regime, in turn became passive objects of transformations occurring within the "subjective" realm. Naturally, the subjective or "software" factor of cultural paradigms that function as action programmes comes to dominate in the short term range of deep historical changes; the objective or "hardware" factor of political institutions, in contrast, affects the shaping of a given political status quo primarily in the long run. It is no coincidence that structuralist, determinist theories of democracy have lately been rejected in favour of activist-possibilist approaches within competitive concepts of democracy.[1]

The software factor can change institutions quickly, but is itself subject to very slow processes of change only. New political culture, however, is not a precondition but itself a result of system change, and materializes only at the very end of the consolidation process. Its emergence represents a fundamental turning point in system change, i.e. the massive appearance of democrats after a long process of democratization. This is what, following the Spanish model, I call the invention of democratic tradition: the establishment of a new civil society imbued with the values of democratic political culture. In this study, I deal with the two major societal conflicts or "value conflicts" that have led to the powerful reemergence of democratic political culture in Hungary and, after the breakthrough of the mid-1990s, into the initial stages of the Spanish pattern of democratic tradition making.

Consensus on fundamental values or "civic culture" cannot be a precondition for systemic change, nor even of democratic transition; rather, it is the result of a process, although we are of course not dealing here with a simple mechanism of cause and effect. Rather than a one way causality, we need to elaborate a dynamic model based on a concept of circular interaction. In 1970, Dankwart Rustow first suggested such a dynamic model by emphasizing the role of political culture as an agent of democratization while anticipating its full conversion only towards the final, "habituation" stage of change. He pointed out that "a dynamic model of the transition must allow for the possibility that different groups – e.g. now the citizens and now the rulers, now the forces in favor of change and now those eager to preserve the past – may furnish the critical impulse toward democracy." New democratic rules need first to be laid down; then "there must be a conscious adaptation of democratic rules", and "finally, both politicians and electorate must be habituated to these rules". It is, after all, the large-scale emergence of democrats which proves a democratization process to have been successful.[2]

I take as point of departure the Rustow model and its three stages of systemic change. In each of these stages, the role of political culture with respect to institutional transformation changes fundamentally and as follows: (1) during a pre-transition crisis, political culture erodes the existing institutional structure and prepares for a new one, (2) throughout democratic transition, institution building takes the lead as the main dynamics of democratization, and (3) in the period of democratic consolidation, political culture again comes to play the decisive role in creating the democratic traditions of a newly organized civil society.

Moulding and modeling

For Latin American and Southern European democratic transitions, this contradiction between institutional discontinuity and cultural continuity has been described as one between frequently abrupt transformation of institutions such as parties, parliaments or social organizations, and the *longue durée* of cultural tradition.[3] My hypothesis is that this contradiction has manifested itself even more clearly in East Central Europe than in those countries, since they had

- the longest and toughest periods of authoritarianism,
- historically, the highest occurrence of re-democratizations,
- the socio-economically most strongly rooted, isolated social and political subcultures,
- the most ossified political institutions, or
- showed cumulative effects of any or all of these factors.

East Central European political systems were stabilized excessively during the long periods of well-entrenched authoritarian regimes. Their abrupt transformation remained inexplicable, as the innovative content of political culture and its role in long-term historical processes was obscured by such institutional over-stabilization. Thus, social movements geared up for revolutionary action by their related political subcultures emerged unexpectedly, seemingly from nowhere. I here develop a fourfold model in order to explain the sequence of system changes in Hungarian history that results from the interaction of subjective and objective moments. The factors are those of (i) political culture, (ii) institutions, (iii) political socialization and (iv) social movements. All these factors are conceived as antipodes of, first, political culture and political socialization as software versus institutions and movements as hardware, each representing an "objectification" or "subjectivization" on the part of the other; second, of political socialization and social movements as objectifications of political institutions and political culture, and new political institutions as objectifications of social movements and political socialization.[4]

System change represents collective action by elite-led or grass root movements in order to alter fundamental political institutions. In this process, long term value orientations of political culture prevail against the short term indoctrination resulting from political socialization. Political culture profoundly reshapes the political landscape through social movements. I thus apply the fourfold model to Hungarian system change between 1988 and 1995, i.e. to the periods of pre-transition crisis and of democratic transition. This model, however, needs to be formulated in three different ways: historically, as a long term determination model; as a meso-model for the analysis of consolidated political regimes; and finally, as a short term determination model when applied to political transformation (see Figures 1–3).[5] This last may be differentiated further by adding an elite-mass dimension that could account for different types of transitions and democracies (see Figure IV) following the Schmitter-Karl model.[6]

This paper therefore deals mainly with political culture and political socialization, i.e. with fragmented, separated, over-competitive, mutually exclusive and enclosed political subcultures, and with simultaneous, inimical and equally antagonistic mechanisms of political socialization. Due to the lack of socio-cultural homogenization, Central European societies have historically been "pluralist" societies whose different strata also generated different types of production and lifestyle, and often shared different ethnic and religious backgrounds. They represent different historical stages, or "alternative pasts" of our region. There is, as my compatriot Karl Mannheim put it, a "synchronity of asynchronity", i.e. a coexistence of structures from different stages of history. Lopsided and belated modernization – industrialization and urbanization – has been unable to transform these societies into unified national structures comparable to those in the West. On the contrary, it has led them into relative isolation and preserved various aspects of half-developed, heterogeneous, capitalist societies. This is a long story which, *mutatis mutandis*, could also be told of other regions and continents, and which offers itself as a historical and theoretical starting point for my analysis.[7]

All of Hungary's system changes have appeared, therefore, as a *Kulturkampf*: a political struggle between isolated subcultures representing conflicting social strata. These strata have existed only and insofar as they have manifested themselves in subcultures – that is, in a way of life, a mode of production, a style of political discourse and a rudimentary set of socio-political activities. The point is that historically, these strata have not had the opportunity to achieve a measure of political organization or properly express their interests. They have remained politically silent and even "dumb", except perhaps for brief periods of manifest crisis and in anomic outbursts of discontent. In their isolation and inadequate articulation of interests, they were ill equipped for learning the ropes of politics and rational compromising, i.e. for "speaking politically". Rather, they remained all but enclosed in their small worlds of isolated subcultures nestling in self contained socio-political universes.

The rare moments of system change, mostly violent revolutions, were short periods of political learning followed by long periods of national and sub-national amnesia. Thus, the critical extent of common understanding has never been reached in political discourse and action. The "language course" always had to begin anew and with fundamental changes in the vocabulary; it ended soon afterwards, and just as abruptly. Even covert and rudimentary forms of political interest representation turned into obscure undercurrents during the long periods of stifling status quo, that is, they seemingly disappeared and then re-emerged from nowhere with a vengeance. Although everyone thought, after the long and hard times of state socialism, that historically rooted subcultures had disappeared, they have merely surfaced and proved to be more lively and active than ever.

Political, or rather socio-economic-political, subcultures are complex phenomena. Three levels or modes of organization may be identified: First, *customs* understood as sets of general behavior patterns of a given class or group, and which operate entirely on an unconscious and metacommunicative level, i.e. which are expressed mainly in emotions, attitudes and prejudices. Second, the *worldview* and its expression and formulation as political discourse, which is a message only members of a particular group may decode, and which therefore serves as everyday vehicle for communication or practical philosophy re-asserting group identity and identification. Finally, the particular competitive *ideologies,* which lend more or less coherent theoretical expression to worldviews and are advanced by representatives of a given social group on the level of national political community, in the public arena, at political manifestations and in the form of protest from abroad.

These three levels or modes have developed very differently in different strata. Some social strata may not even have ideologies at all, or only in a very rudimentary form. We can now classify the remaining factors of the fourfold model in accordance with these levels of political subculture as those of political institutionalizations (from grass root associations to parties), political socialization mechanisms (from families to universities) and social movements (from protest behavior to organized movements). All forms found at present, however, are part of (value-based and over-ideologized) "cultural politics", rather than (interest-based and rationally argued) "political politics" prevailing in mature democracies. The present stage of Hungary's democratic transition may thus be characterized primarily as one of transition between cultural politics and political politics. It is generally agreed that Hungarian political parties have grown out of "tribal" subcultures; now we are witnessing the emergence of their internal-external rearrangements in accordance with fully developed social cleavages in a process blueprinted in the Lipset-Rokkan model.[8]

The "long value conflict" and awkward re-Westernization

Hungary's soft dictatorship allowed for soft societal resistance and the cautious revival of European or Western values against official ones. Public life was split between official and informal public spheres. A strange "peaceful coexistence" and complementarity on the one hand, and a fierce, protracted and tortuous value conflict on the other reigned between them. The latter led to a continuous, if cumbersome and contradictory, Westernization of Hungary. Western European values won out step by step during the 1980s, and were finally accepted publicly even by leading technocrats and most of the political elite, albeit with a degree of cynicism. The value conflict opposed paternalism-infantilism and pragmatism-individualism, and was lost by the state socialist regime long before its political collapse. "Returning to Europe" became a widely and openly accepted agenda in 1988–89, and was adopted as the Németh government's political philosophy prior to the first free elections held in the Spring of 1990.

At first, the "long value conflict" brought with it a return to Hungarian traditions in political discourse, i.e. to its "segmented market" of political ideas and mentalities.

Hungarian society (as that of any East Central European country) has traditionally comprised the following five major macro-groups:

1. the "post-estates" (or gentroid) political class
2. urban middle classes
3. working classes
4. rural or peasant strata
5. the marginalized as a subclass.

These macro-groups have engendered political subcultures which may be characterized mostly as civilizational-cultural complexes with their own ways of life and modes of production. All have survived state socialism in isolation, although they either substantially declined in importance (1,2,4,5) or grew markedly (3). Most significantly, fundamental values and lifestyles of these groups have survived – at times in obscure or roundabout ways as networks of "good families" (1) or in the form of competitive individualism (2), in the circumscribed existence of old state bureaucracies and the legal profession, or in that of commercial professions. The working classes (3) never felt like a new ruling class, and opposed the new political elite even though it had emerged mostly from their ranks. Their existence was partly traditional and partly urbanized, and largely defined by the lopsided and belated modernization of the early 20th century. Hungary's rural strata (4) were mostly modernized, the majority of them having given up traditional peasant existence and developed a complex web of activities that included trade and services performed in a quasi-commercial spirit. Rural life, however, retained a stigma of backwardness mainly as a result of missing infrastructural developments.

The poor (5) officially disappeared, and their situation did in fact change drastically in the premature or pseudo-welfare state. But the core group remained unaffected by these improvements, and during the 1980s the poor returned in growing numbers. In the 1970s, they were rediscovered and identified by critical social scientists as "cumulatively disadvantaged strata". Of late, they have been particularly affected by the interaction of economic, cultural, and political factors working to their disadvantage. In the 1970s, major trends toward vertical social mobility slowed down, came to a halt and were finally reversed by increasing internal reproduction of groups. A new and very broadly based intelligentsia, however, grew out of the first four groups (1,2,3,4) in the decades of state socialism. It was a heterogeneous "superclass" with large overlaps in all four cases, which was imposed on the entire social structure and topped only by the political elite. But in Hungary, this elite spread beyond apparatchiks and became increasingly differentiated according to political, economic and cultural roles. It was confronted with an increasing demand for its expertise, leading to overlaps with the intelligentsia "from above".

This brief outline of the major social strata – diverging from Western social structures in characteristic ways – now allows me to embark on a description of social movements. I take social movements to be mobilizations of social strata on the basis of their set of values and against any *de facto* socio-political order, that is, against its institutions and political socialization mechanisms. Western conventional wisdom has it that during the periods of pre-transition crisis and democratic transition, ECE countries knew only elite struggles, but no social movements – except for Polish Solidarity. Contrary to this assumption, Hungary's social movements have been extraordinarily diverse and active in recent decades, since civil society was rallied vehemently along the cultural faultlines described above. It was a mobilization against the political regime, even if these movements made no direct or overt political claims but confronted the regime "culturally".

Economic resistance of a shadow economy, as well as socio-cultural resistance of a shadow society, were manifestations of the value conflict between Western modernization hampered by traditionalisms on the one hand, and aggressively imposed Eastern variants of modernization on the other. The deep structures of Hungarian society (with their positive as well as negative features) resisted the dual onslaught of alien values, first passively, then actively, and finally by producing a range of hybrid forms mediating between Western and Eastern patterns. At the same time, this process of awkward re-Westernization also challenged any palpable modernization within the "walls" or limitations of the Soviet's external empire. It is a very controversial, though mainly progressive, inheritance of the present democratic transition, unique even for East Central Europe insofar as it constituted a quasi organic postwar modernization in the wake of "anti-democratic" Stalinist transition.[9]

The long conflict of values was led and directed from above by the reform minded intelligentsia. Its soft resistance against soft dictatorship broadened into a vast but loose reform movement supported by large parts of the population. An obscure but very influential reform movement, it rested on the confluence of economic and socio-cultural resistance. It entered the political sphere, however, only in 1988. Overt political confrontation also brought with it a growing differentiation of, and confrontation within, the reform movement. In 1988, all major strata produced socio-political movements in a "cultural" form of their own, which soon reorganized as political parties.

Political socialization mechanisms also changed, in my opinion, in accordance with the stages and forms of these social movements. The conflict of values gradually delegitimized and eroded official political socialization mechanisms, and facilitated the emergence of new ones reasserting primary socialization stages located within the private worlds of family and civil society, as against schooling and the formalized ideology of public life or media as secondary and tertiary stages of a socialization that were monopolized by state socialism. But although the regime's attempts at indoctrination and homogenization of values failed conspicuously in the 1980s, traditional paternalist-infantilist attitudes of Hungarians were nonetheless reinforced considerably. Yet, traditional capacity for indirect political resistance was strengthened equally by these moves; the formal structure of official pressure became increasingly self-defeating when applied to the informal structures of civil society and social movements. The first value conflict had thus been lost by the state and won by civil society in Hungary before the actual political struggle of system change even began. Pre-transition crisis − the first period of system change − occurred in 1988, when this conflict of values had already been lost by state socialism and its defeat was recognized by Hungary's political elite. Recognition of defeat led this elite to capitulate in 1989, well in advance of state socialism's general collapse in Central and Eastern Europe.[10]

Value conflicts and political subcultures

The first conflict of values gradually ended in 1989. The victory of European values in political life, however, proved premature, as another such conflict esclalated just after transfer of power early in 1990. Thus, pre-election cultural struggle among a broad range of parties was subsequently cast into a dichotomous antagonism between Traditionalizers and Europeanizers, i.e. between those whose priority was "Hungarianness", and those who favoured Europeanization. This new battle over values essentially ended with the electoral defeat of neo-traditionalist conservatism in Spring 1994, although the latter is a mindset still prevalent in some social strata. Before we can analyze the stages of this second conflict of values, however, we need to examine in more detail the traditional political subcultures which can be identified as protagonists in this struggle.

The most important factor in Hungarian political culture and its systemic dynamics during the late 1980s was the persistence of inimical and fragmented political subcultures. Lopsided state socialist modernization had – to some extent – changed but not eroded the ways of life of major social strata, at least not insofar as they were manifest in customs. On the contrary, *customs* gained in importance as expressions of group identity to the extent that more confidently, cogently verbalized forms of such expression and interest representation had but limited opportunity to develop. This zero-level group identity was the source of diverging and invigorated political *discourses* in the late eighties, when political crisis allowed for articulation of dissent in a more overt, adequate and directly political manner. My interpretation will thus focus on this level of "worldviews", on the media as message, on form rather than on content, or simply on how something was said rather than on what was being said. These conflicting discourses later became in turn points of departure for the particular competitive *ideologies* of socio-political actors and/or parties.

Customs are alternative socio-economic lifestyles. The actual process of framing – a semi-conscious formulation of interpretation schemes or models through which individuals of a given group can come to terms with major social and political events – starts at the level of discourses embodying worldviews. Through the framing process, major social changes are approached with a common conceptual framework of interpretation in turn set within a particular cultural field of meanings. Solving problems, conflicts, or any confrontation is thus given a general direction for a particular community. At the same time, such frames are the conceptual foundations of subcultures situated on a higher level, of, that is, party programmes and ideologies as consciously deployed framing strategies in the field of interaction between public-political life and the media.

My hypothesis is that in Hungary during the late 1980s, two basic sets of values emerged as alternative reactions to the (imminent or predictable, then actual) collapse of state socialism. The traditionalist value system hearkened back to pre-socialist history (late nineteenth century to interwar period) and emphasized traditional values such as "God, Nation, Family" (as cast into one party slogan), "customs", or, still more simply and lambently, "Wine, wheat and peace". In contrast, the modernizers expressed the urgent need to "Return to Europe", with mandatory acceptance of the most recent, most modern, European values. These scenarios of Returning to the Past and Returning to Europe have given divergent responses to a common question: where to start again and whither to go after fifty years. The first of these responses built upon an internal logic of national history and its particularities – on eleven centuries of Hungarian continuity – in the spirit of return to a Golden Age That Never Was. The second also turned to national history for inspiration, but only to those repeated efforts at modernizing and catching up with Western Europe which had been made in the – usually naive – belief in Western assistance and in unique opportunities presented by a given historical situation. ,

133

It is easy to identify two "camps" (simlar to Austria's two opposed *Lager*) behind these values or fundamental worldviews. Hungarian public opinion surveys have shown that the six identifiable basic values may easily be divided into two sets of three strongly correlating values. Both groups of values form networks or schemes pervading the history of the recent system change. There has been a close correlation between "national-minded", "believer" and "conservative" attitudes of the right wing value triangle, and between "left-oriented", "democrat" and "reformer" attitudes of the left wing set. A somewhat weaker affinity between the first triangle and a "law and order" orientation, as well as one between the second and "libertarian" thinking, has been found. More and more Hungarians, at present 82%, are able to place themselves on the Left-Right axis, and an average score of 4.7 on a scale of 0 to 10 suggests that the majority is slightly closer to the Left end of the spectrum.[11]

These two basic approaches or value sets first enter political discourse as symbols, meanings and sentiments, and subsequently appear in competitive ideologies as concepts and theoretical systems. "Nation"-centered and "European"-centered approaches are both exclusivist, but not ecxlusive, that is, there have been overlaps despite all claims and efforts to the contrary. In the shaded areas between the opposite camps, there have in actual fact been many nuances in the value hierarchy of Hungarianness and Europeanness. Yet, the major front lines in the second major value struggle corresponded more or less to those between the three governing and the three opposition parties in the Hungarian parliament (1990–4). The growing acceptance of European values (about 90% of Hungarians support the adoption of Western European political models for Hungary) may thus indicate a trend of Europeanization also within Hungary's party system.

Historically, there have been four different strands of "traditionalist" discourse, and only one central discourse for "modernizers" that has recently begun to desintegrate into different sub-discourses under the pressure of competitive party ideologies. Thus, *legalist-paternalist* political discourse uses the vocabulary and symbols of the traditional political class, a language traditionally expressing an exlusive claim to power and political leadership. This has been the discourse of a minority which rarely addressed, or was understood by, a large public. It has always been an esoteric language of rulers and served to emphasize their distance from the ruled. It has legitimized the rulers' monopoly of power by reference to historical continuity, expertise and inert compliance on the part of the population. This has historically been accompanied by *nationalist-populist* political discourse of a lower nobility expressing their claims and remonstrances on behalf of the "people" and "nation" against the high world of politics. They resorted to a separate language of political "kindergarten" voicing their particular interests. Populism also grew out of this tradition. It acquired a more and more rural basis, with (traditional) peasantry as an imaginary core of the "genuine" people. Thus the style of populism increasingly turned towards the archaic, provincialist and regressive, although it has always played a positive role by countervailing exclusivist high politics.

Christian-Religious political discourse has complemented, rather than opposed, "official" and legalist language. Its messages have been decidedly simplistic, though with significant variations according to whether the middle classes, the peasantry, or uneducated elderly people were to be taught the necessity of submission to, and compliance with, traditional authorities. Out of this framework, however, grew a *literary, artistic political* discourse and milieu that powerfully opposed secular and ecclesiastic authorities. Its moralism became a substitute for the missing democratic legitimation of authoritarian regimes. Poets, writers, historians and other public figures felt that they represented the people or nation by having won public acclaim in their arts or professions, and by maintaining moral standards against authorities in the face of dire repression.

As against this discourse and public role of poets and writers as fathers of the nation, all others are closely tied aspects of pre-modern, pre-democratic politics. The decline of literary discourse during the 1980s demonstrated the limitations imposed by such pre-modern politics. Although literary language could still reach large audiences, it proved less and less capable of articulating conditions and pressing demands of a complicated modern world. Thus, most public figures who could claim a central role in non-democratic, non-representative politics became disillusioned with their status of moral prophets of the nation, and realigned with the traditionalism espoused by the Antall-Boross regime.

The modernizing discourse, finally, has generally been a *European-Rationalist* one (re-)emerging and dominating as a worldview over pre-modern and state-socialist discourses as a result of both the first value conflict and awkward Westernization. It has also come to stand for the quasi-organic Hungarian modernization and its continuity in our own decade. Therefore, it has been attacked as "judeo-liberal-communist" by traditionalist discourses, in extreme right wing versions even as a conspiracy against true Hungarians and their historical values. As point of departure for framing strategies, we can focus on the correspondence between traditionalist discourses and right wing party coalitions. Here, not political programs but common ideology served as the basis for forming such a coalition. Thus the Christian Democratic People's Party (CDPP) is an overtly ideological party relying exclusively on religious discourse. The Independent Smallholders' Party (ISP), too, is a "single issue" organization, addressing itself to the most traditionally minded layer of peasantry. Its language is simplistic and populist, or, more accurately, it accepts this mentality as power base. The Hungarian Democratic Forum (HDF) was launched in the spirit of literary discourse before this style came to be marginalized. In HDF the divide and confrontation between party and movement was expressed in the fight of the legalistic-paternalistic and nationalist-populist discourses.

These traditionalist discourses correspond more or less with "pre-modern" social strata, while the "European" discourse has its roots in the lopsidedly modernized strata. The former have increasingly come to represent the mentality of losers in system change (in the long run, even the traditionalist-conservative elite will be threatened by the process of Europeanization). The latter has been that of (potential) winners who may have acquired some European or reform impulses or accumulated skills, knowledge, networks and even some property. There has been little in the way of dialogue among major political discourses or competitive ideologies, nor even among party programmes. Political life and public opinion are highly fragmented, and discourses are exclusivist. They have not so much tried to argue, as to to dominate and force their rivals out of public life. As party conflicts within the (1990–1994) coalition demonstrated, this was true not only for the two camps but to some extent also for the competitive traditionalist discourses. All discourses have tried to appropriate, own and monopolize the media, some from within, others from without.

Empirical data covering the late 1980s clearly indicate the clash of these two camps and their particular discourses, as well as of their respective socio-political environments and actors. When interpreted formalistically, however, these data can be seriously misleading in any comparison of Eastern and Western, or East Central European, political cultures. The meaning of politics and the role of political culture differs from time to time as well as from country to country; we need to analyze the particular conditions in each ECE country, namely, in this paper, the Hungarian peculiarity of a conflict between Europeanizers and Traditionalizers.

The *Political Yearbook of Hungary* (edited by the Hungarian Institute for the Study of Democracy) has recently published the results of public opinion surveys conducted by Central and East European Eurobarometer (CEEB) and by Modus Institute in Hungary, on democratization and marketization. Comparative data would suggest that, as a result of the "authoritarian renewal" during the early 1990s, Hungarians' decline of interest in politics has been the most dramatic of all Central and Eastern Europe countries: 57% as against Poland (48%) or the former Czechoslovakia (31%). Regional comparison of this kind, however, is seriously misleading since the meaning and function of Hungarian politics have changed drastically during the late 1980s and early 1990s. Here, the most radical democratizations took place formally or legally and in macropolitical institutions. Thus the loss of interest, though undeniable, has expressed a great deal of disappointment with the "new" politics against the backdrop of previously formed high expectations by politically active sectors of the public. In Hungary, political mobilization happened earlier and reached deeper in its forming a multiparty system with a new democratic constitution already in the late eighties. Disappointment was, of course, greatest in the region as a result of the poor performance of parties and parliament, and as Hungary was entering a demobilization phase. But disappointment of this kind is surely a symptom of frustrated agility rather than apathy or disinterst in matters political.[12]

Historically, flagging interest in politics may be understood as an accumulation of bad experience. In Hungary, a sizeable nobility constituted the political nation and, as part of the lopsided modernization during the late nineteenth century, transformed itself into a political class. The restriction of political deliberation to a small class or "caste" was certainly a heritage weighing heavily on Hungarian society. Although the monopoly of the political class was also subject to gradual erosion, it was broken only during the short periods of revolution. A sequence of thwarted revolutions between 1848 and 1956, moreover, entrenched a certain sense of despondency with regard to political commitment. To get involved in politics came to be seen as dangerous and futile, not least since the West would never really support substantial democratization in Hungary. Nonetheless, the general lesson Hungarians seem to have drawn from history has not so much been one of political apathy, as the need to create more indirect forms of resistance to politics and politicians. Civil society movements in state socialism showed classic features of this indirect or soft challenge to the regime. The power of refusal wielded by the masses and their associations effectively thwarted realization of government policies, and frustrated leaders. In this vein, Hungarian political culture may be described as a kind of sober realism, pragmatism and cautious optimism (although the latter sometimes appears as a pessimism resulting from unrealistic expectations). Fanaticism and extremism have accordingly been marginalized and have so far failed to attract significant responses from larger segments of the public.

This hard core of soft resistance turned actively political in the late 1980s. But the Hungarian public was soon frustrated anew by the new democracy's low efficiency and by the poor performance of political leaders under the national-conservative government. Neither problem solving capacity nor popular participation improved significantly in the early democratization process, and actually worsened in the eyes of the public with respect to certain issues in the political field. The distance between people and politicians grew considerably, though obviously not because politicians were thought to have been more competent before 1988, but because expectations had risen dramatically in the meantime. Although an overwhelming majority of Hungarians believed that democracy was generally more efficient than authoritarian government – 93% in matters of education, 88% on economic policy, 87% on issues of poverty, etc. – they did not feel that the first post-authoritarian government could realize this potential. On the contrary: in 1989, at the very beginning of democratization, the majority of respondents (57%) expected problem solving to improve with democratization, while 33% thought it would remain unchanged and only 10% believed it would decline. But by 1992, only 29% of respondents had retained an optimistic outlook, while a relatively stable 34% expected no changes and a drastically increased 37% felt that efficiency had actually deteriorated.

This widespread disappointment was, however, directed primarily against the neoconservative goverment rather than against democracy as such. Popularity of that government declined steadily as a result of its poor record of Europeanization, i.e. of marketization, democratization, privatization and pluralization. Already in 1989, an overwhelming majority of Hungarians opted for a Western European paradigm of development (88%). This figure has increased since then (93% in 1992), making Europeanization an all but hegemonic value and the broadest base for consensus in Hungary. Politically, the matter is quite clear: the test to be passed by any party or politician has definitively been couched in the discourse of Europeanization.

Early senility of a young democracy: neotraditionalist anti-Europeanism

The first democratically elected government failed to meet the public's high expectations. It proved incapable of continuing the Europanization trend in an accelerated and adroit manner and under radically altered, i.e. radically improved, circumstances. The value conflict between Europeanizers and Traditionalizers was thus, however temporarily, lost by modernizers in 1990 on a political level and resulted in an "authoritarian renewal". But traditionalizers dominated only the first government and its coalition partners, and Hungarian society's immune defense soon made it plain that not the conflict, but only a first struggle was over.

Comparative data on Central and Eastern European countries show that a majority supports Europeanization through joining the European Union. Such positive responses are usually burdened, however, by severe cases of "cognitive dissonance" (e.g. in favour of Europeanization but against marketization, etc.). In any ECE country, a majority of the public typically endorses Europeanization but (1) holds unrealistic views regarding the country's prospects for joining and flourishing in the Union (2) has developed exaggerated expectations concerning EU assistance and the short term consequences of membership, and finally (3) does not have a clear picture of either the prerequisites for Europeanization or the tremendous social price to be paid for it. However, for Hungary the crisis of disillusionment in the face of harsh realities may now be said to have been overcome, and major cognitive dissonance seems to have almost vanished.[13]

There is, at present, no room for an openly anti-European party in Hungary (except for tiny groups, and despite the fact that the Party for Hungarian Justice and Life, or the

137

Workers Party are not immune to anti-European stances). Those who think that Hungarian political culture or basic value sets are not sufficiently European may well be surprised by data illustrating value priorities in Hungarian society. After two or three decades of awkward Westernization, Hungarian's value orientations are even somewhat excessively modernized, since priorities accorded to competitive individualism, achievement orientation, pragmatic intellectualism and individual self-reliance surpass the possibilities of their realization.

Former opposition party supporters (Hungarian Socialist Party, HSP, Alliance of Free Democrats, AFD, and Alliance of Young Democrats, Fidesz) generally stood by European values more than supporters of the government. But there is a continuum of values and mixed priorities rather than a chasm between them, since most people uphold European values to a great extent anyway. Traditionalizers prevailed only in the new governing elite on the one hand, and among extreme populists and their followers on the other. Their mentality evidently is archaic and provincialist. They may well embrace a Europe of medieval Christianity or of the nineteenth century, but their agendas are not much situated in the twentieth. Obviously, their outdated ideas, in conjunction with stalled Europeanization in economy and polity, were major causes of the government's, and the entire regime's, declining popularity.[14]

The short-lived triumph of an elite over the long term Europeanization trend in Hungarian political culture pitted politics against society, and government against the public. This in turn led to changes in the government's and ruling party's policies, both of which eventually – though half-heartedly – turned against extremists, anti-Europeans, provincialists, and populists within their own ranks. Internal struggle over direction intensified, in particular, within the HDF in Spring 1993. But if we are to find an explanation for the emerging anti-Europeanism of neoconservative parties, it will be necessary to attempt a periodization of social movements and parties involved in system change.

There have so far been four stages in the political or party history of system change. Actors (social movements and parties) as well as institutions and political socialization mechanisms have changed beyond recognition at each of the following stages in party transformation:

1. Emergence of a large number of social movements organized as quasi-, "embryonic" or pre-parties between May 1988 (the date of a party conference and manifestation of crisis) and June 1989 (the start of round table negotiations and indirect legitimation of democratic actors),
2. transformation of pre-parties into formally organized, early or "infant" parties, and reordering of the entire political scene prior to the Spring elections 1990,
3. parliamentarization of major parties after the elections, and power transfer between governments and ruling elites, up to and until the major crisis of the new goverment in Autumn 1992,
4. appearance of an urgent need for social pacts and a "public policy" approach of legitimizing and institutionalizing organized interests in a tripartite setup, beginning in 1992 and extending to the present day.

The last two stages have also been major periods of value conflict between Traditionalizers and Europeanizers, which have been carried out in the media, by means of intra- and interparty infighting, and in a struggle against re-emerging "oppositional" socio-political movements. As we have seen, the end of this first value conflict also marked the beginning of system change, while the second conflict occured only during the third stage. In 1989, during a transition period between the two value conflicts, a consensus including reform socialists, social democrats and various opposition groups emerged on the basis of a "Return to Europe" programme. This was a phase of homogenization with views and programmes converging towards a general, rather abstract and unspecified acceptance of parliamentary democracy, rule of law, market economy and political pluralism as frames of worldview common to all political groups. The acceptance of a Western European type of constitution by all the participants in national round table talks was the most striking illustration of this, as was the preoccupation with legislative steps toward creating a market economy in parliament. Both efforts led to profound institutional changes and laid the bases for a thorough reshaping of macro-political institutions.

In 1989, oppositional movements were still weak, and competition among them was rather limited. Their particular profiles were relatively hazy, at least in relation to the overwhelmingly cherished European values. The only remaining opponents to Europeanization were not to be found among the declining old party elite, but among the new literary movement within the Hungarian Democratic Forum which advocated a "third route" between Eastern and Western paradigms. That concept was, at the time, still quite open and flexible; only later it degenerated into nationalist populism. The narrow-minded "historical" parties (ISP and CDPP) were just entering the political scene, but their ignorance of contemporary Europe initially prevented them from acquiring political momentum.[15]

The second, transition stage of party transformation saw the victory of subcultures that first organized as social movements, and later as new political parties, around their specific political discourses. Party programmes were similar in what they said, but differed significantly in how they said it, in, that is, their styles of articulation. The *Kulturkampf* of political discourses began hesitantly in 1989, and was rife from roughly the beginning of 1990. It followed the value conflict, since it involved all parties in a ubiquitous *bellum omnia contra omnes* of subcultures prior to and after the elections. The weaker "infant" parties fought very aggressively for their own identity in this precarious and impenetrable "hundred party system". After the collapse of the old regime, and accompanying the spectacular changes in global diplomacy, the entire political style changed beyond recognition. It became aggressive and overly competitive. All parties claimed exclusive representation of previously quasi-consensual European values in this phase of heterogenization – denouncing each other as anti-European and/or anti-Hungarian – but failed to create distinct images of Europeanness. In the crude and heavily ideological election campaign of 1990, party elites routinely displayed a thoroughgoing amnesia concerning earlier periods of, however awkward, Westernization. In order to win political prestige, most tried to present themselves as resistance heroes in harshly worded and intellectually undemanding discussions of the old regime.

The main change was the Hungarian Democratic Forum's development from a loosely aggregated popular movement to a tightly organized party under new leaders coming from the traditional political class. These professionals replaced populist writers – possibly discredited as leftists, but at least showing some sensitivity to social issues – from ruling positions, a take over which forced "movementists" and populists into an internal party opposition. The main cultural and political conflict, however, was to be the one between moderate centrists with their legalist discourse on the one hand, and extremist radicals drifting towards a right-wing, nationalist and populist discourse on the other. They emerged only during the third stage of party development. Before the elections, the fiercest confrontation included mutual accusations and symptoms of extreme cultural tribalism, and took place between AFD and HDF. It was during this campaign that Europeanness was first, though indirectly, criticized as "cosmopolitanism". After Autumn 1989 and for some months to follow, HSP maintained popularity levels with its modernization-cum-Europeanization program, but the new slogan soon began to take effect. That is, more and more people saw a possible winner in HDF and wanted to join, leaving a "spiral of silence" surrounding HSP. The general situation in Spring 1990, more particularly the fact that the election was mainly a referendum against the old regime, distorted the emerging party structure considerably. Citizens' value structures generally did not correspond with that of parties in parliament. In particular, the European values of a sizable majority clashed with those of the new traditional-conservative governing elite. This non-correspondence inevitably and quickly led the new traditionalist, right wing coalition government into a legitimation crisis.[16]

Europeanness was not an election issue in 1990, since Returning to Europe was, as an abstract formula, widely accepted anyway. This is also why Europe was no concern for content analyses of party programs. But Europe became the major issue immediately after the elections, that is, it marked the principal cultural-political cleavage between government and opposition. *Kulturkampf* thus defines the third stage in the development of Hungarian parties. Only in 1990, parties began to offer divergent images of Europe which were derived mainly, although not exclusively, from their liberal, socialist and conservative pan-European affiliations. Hungarian parties, however, are still at pains to integrate into these international organizations.

After the 1990 elections, system change should have gained new momentum qualitatively and quantitatively, i.e. both in substance and in extension. Political actors in power attempted to transform their programs into social movements organized around particular political subcultures. Small coalition parties turned into single issue anti-European political actors: the ISP took an active interest only in the reconstruction of traditional peasant farming, while the CDPP devoted itself exclusively to the issue of reprivatizing former Church property. In the service of such nineteenth century ideals, these parties have repeatedly blocked all other system change legislation. The CDPP fiercely opposed any separation of Church and state, declared itself a Christian party, and rejected ideological neutrality in education. While the ISP wanted to restore the monopoly of traditional peasant life in the countryside, the CDPP tried to impose an ideological monopoly of the (Catholic) Church. It was because of these programs that the value conflict broke out early in 1990. Since then contemporary Western Europe, and its structures have become the basic frame of reference for political debate. The political struggle has not only concerned Europe, but has been one in defense of Europe, i.e. of European integration versus narrow minded Hungarian provincialism.

Anti-Europeanism of the HDF has been more complicated, the "Hungarian Road" movement within this party (which later reorganized as Party of Hungarian Justice and Life) only covering part of the whole story. The center and liberal wings of HDF by no means neglected to deal with the EU issue or to advance their own images of Europeanness. But their backward looking, outdated mentality, as well as their inadequate understanding of contemporary Western Europe, turned into a major problem. Both became a burden for the new traditionalist-conservative governing political elite, which pursued a contradictory strategy of "archaic" pro-European rhetoric and anti-European political practice. The Antall-Boross government and its coalition parties generally accepted the need for Europeanization and structural adjustment to the EU, but their political decisions were in many cases inconsistent with this stance. This was evident from policies regarding the Church, Privatization or local government legislation, or, most markedly, in the rejection of organized interests as socio-political partners in transition. Half-hearted legitimation of socio-political actors was the salient characteristic of the early fourth stage, when Hungary's first and highly controversial social pact was concluded in February 1993.[17]

Democratizing political culture: dangers and opportunities

Both the images of Europe created by parties and other political and socio-economic actors, and the *de facto* contradictory Hungarian Europeanization policies of the mid-nineties would deserve further in-depth analysis. In conclusion to my discussion of Hungarian political culture and its development from the late eighties to the mid-nineties, we may say that this transformation has seen the victory of political subcultures over previous institutions, and led to the establishment of new democratic institutions. Europeanization as a core value has become the major social and political program for Hungary in the nineties, but temporary dominance of Traditionalizers over Europeanizers among political elites created adverse conditions for its implementation. The Hungarian public still supports Europeanization, but that support has suffered from the EU's benign indifference and from controversial policies of the former government. As a result, it is now more hesitant and diffuse than it was only a few years ago.[18] In the early 1990s, Hungarians proved increasingly dissatisfied with the development of democracy and the performance of the former government, a discontent reflected in the results of the 1994 elections. The new government, however, has also given a new impulse to the democratization of political culture and the Europeanization of the Hungarian polity.[19]

That government, however, created new major controversies during its first year in office. The harsh economic austerity measures which had forever been postponed by the previous government have now come into effect, and have fundamentally divided public opinion. In this way, the positive turn toward Europeanization, democratization and marketization has in part been offset by the negative reverberations this has had on popular mood. One result of the new social-liberal government's assertive economic crisis management has been a renewed susceptibility to the lure of social demagogues, rather than to the kind of nationalist clamour prevailing at the beginning of our decade. Populism has thus reemerged as socially aware populism; like its earlier nationalist variant, it is a disruptive force in the consolidation of democracy.

On the whole, the invention of democratic tradition in Hungary is well under way. But it will succeed only to the extent that it achieves a simultaneous breakthrough in economic consolidation and European integration. No masses of ready-made democrats may today be found in Hungary, but a change for the better has been felt with the accumulation of political experience, and Europeanizers will undoubtedly prevail in the value conflict. Hungarian democratic tradition and political culture have already survived two tough tests in the form of such value conflicts: successful opposition against state socialism and its authoritarian regime before 1989, and resistance against the "authoritarian renewal" led by a national-conservative government (1990–4). On this firm basis, political culture in Hungary is more than likely to achieve democratic consolidation, as well as the definitive invention of its own democratic tradition, within the next decade.[20]

Notes

1 I try to concentrate here on the interaction of subjective and objective factors in the process of system change, and must leave aside related problems. The distinction between short term, politically centred and long term socio-economic determination centred approaches has been described by Pridham (1990, 8).

2 I follow here the concept and terminology of Victor Pérez-Diaz (1990), a paper widely noted and available in Hungarian translation. Dankwart Rustow originally introduced the three-stage model of system change, the third stage being conceived as a "habituation" phase. Rustow also first distinguished clearly between structural-functional analyses of mature democracies and genetic approaches to newly emerging democracies. He indicated, in contrast to received opinion, that the appearance of democrats was a result, rather than a precondition of democratization: "Many of the current theories about democracy seem to imply that to promote democracy you must first foster democrats (...) Instead, we should allow for the possibility that circumstances may force, trick, lure or cajole non-democrats into democratic behavior" (Rustow 1970, 344–5, 345 and 361).

3 See Maurizio Cotta (1991) for a general model of re-democratizations. Pridham (1991) takes up and develops further Juan Linz's distinction between cultural continuity and institutional discontinuity. He also points out that similarities are closer and comparison is more fertile between these two parts of Europe than between Latin America (LA) and Southern Europe (SE).

4 I have described several system changes (with different strata, lifestyles and discourses) in Hungarian history based on the fourfold model (Ágh 1989, 75). The lifestyles and discourses in fragmented Central European societies have been characterized very well by George Schöpflin (1990). He argues that "[s]ocial mobility was low to very low (...). At this point, ethno-national and social cleavages could coincide".

5 (1), (2), (3) and (4) in figures 1–3 indicate the importance of a given factor for social determination, (1) and (2) drawing a first or main axis, (1) and (3) a secondary one, (1) and (4) the weakest.

6 Figure IV is adapted from Philippe C. Schmitter and Terry Karl (1992, cf. also Schmitter and Karl 1991).

7 The theory of "plural" societies has been elaborated in studies on the "Third World", including their cultural consequences or manifestations. This literature has also been extremely fertile with respect to Central Europe, which I have discussed in the book mentioned above.

8 Hungarian political science has produced a wealth of studies on this cultural struggle among parties. I have characterized the recent Central European party system from such a "cultural" point of view (1994), using the same periodization as in the present article.

9 For details, see the work of Elemér Hankiss (1990).

10 The period prior to 1989 has been analyzed in a volume edited by Csepeli, Kéri and Stump (1991). See also *State and Citizen* and *From Subject to Citizen* (1993 and 1994) by the same editors.

11 On political camps, see R. Angelusz and R. Tardos (1992).

12 The results of Eurobarometer (CEEB) have been published in the 1993, 1994 and 1995 volumes of the *Political Yearbook of Hungary*, edited by the Hungarian Centre for the Study of Democracy. Annual summaries of public opinion surveys have been published since the first volume of the *Yearbook* (1988). Rather than quoting the abundant data from this source, I will instead refer only to the work of László Bruszt and János Simon on political behaviour in Hungary. See their summary, "After Antall, before the elections – or 'our democracy and our parties' through the eyes of citizens" (*Yearbook* 1994), from where I take the following figures.

13 The particular contradiction between values of liberty and excessive reliance on the state for social security appears everywhere in the region. Hungary is not an exception, but a less severe case. On vestiges of parternalism and its incompatibility with individual liberties see Bruszt (1989). On

142

Europeanization of the Hungarian value system (individualism, entrepreneurship and pluralism) until the late 1980s, see the analyses and data of Elemér Hankiss (1989 and 1990).

14 Some national-conservative government ministers repeatedly stated that "our Europeanness is in our Christianity", and that "we Hungarians were more Christian than the other European nations".

15 Public opinion surveys in 1988–89 concentrated on comparing Hungary and some Western European countries with respect to some major issues such as market economy and political pluralism. Hungarians showed a consistent preference for Western European models. Empirical data indicate that the average citizen could not discern major differences among the emerging parties (except for their elitist character, i.e. where organized and led by a small intelligentsia). 90% of respondents made their party preference on the basis of "emotions" which correspond, in fact, with familiar political discourses. See Part III of the *Political Yearbook of Hungary*, 1990.

16 On the basis of a large data sample, R. Angelusz and R. Tardos (1991) show that value dimensions were very important in the first free elections, especially for the two small parties (ISP, CDPP). These value orientations formed the profiles of emerging parties. I find particularly interesting their analysis of political and value divergencies in "cores" and "rings" of parties.

17 Images of Europeanness in Hungarian parties have been described in 'The Europeanization of the Hungarian polity' by Attila Ágh, László Szarvas and László Vass, in: Ágh and Kurtán (1995). For institutional development see also Ágh (1994). For details on political culture see Ilonszki and Kurtan(1992) and Plasser and Ulram (1992).

18 Only 37% of Hungarians had a favourable opinion on the EU according to the 1992 survey (45% in 1991), although only a small percentage held a negative opinion, and the remainder adopted a "wait and see" position. Eurobarometer 1992 shows that Albanians, Romanians, and Bulgarians had the highest opinion on the EU because, in my view, they had had almost no contact with it by then and were therefore overly optimistic (see the 1993 volume of *Political Yearbook of Hungary*).

19 The overwhelming majority of Hungarians became increasingly unsatisfied with the national-conservative government (see *Political Yearbook of Hungary* 1995, 595):

Are you satisfied with the development of democracy in Hungary?

	1991	1992	1993	1994
very satisfied	2,3	2,0	2,3	1,3
rather satisfied	28,2	20,0	18,0	21,9
not satisfied	39,2	44,0	44,2	41,7
very unsatisfied	21,0	28,0	30,0	23,9
don't know	9,2	6,0	5,5	11,2

These percentages anticipated the results of the 1994 Spring election to a large extent, since some 20% voted for right wing coalition parties. Public opinion polls in June 1994 indicated that a majority of the population was optimistic about the future of democratic development under the new social-liberal coalition. Specifically, about 50% were optimistic, 29% cautiously optimistic and only 21% rather pessimistic. This survey again reflects election results, in that the share of supporters of the nationalist, right wing parties has oscillated around the 20% mark (see Hungarian Gallup Institute, published in *Magyar Hirlap*, Budapest daily, 6 June 1994).

20 See Ágh (1994). In the present paper, I have focused on these developments in Hungary and on some theoretical issues, since it is too early to discuss the new period of the socio-liberal government in more detail. On specifically Central European cultural traditions see Ágh (1994).

References

Ágh, A. (1989), *Self-Regulating Society*, Kossuth, Budapest.

Ágh, A. (1994), 'The Hungarian Party System and Party Theory in the Transition of Central Europe', *Journal of Theoretical Politics*, vol. 6, no. 2, April.

Ágh, A. (ed.)(1994), *The Emergence of East Central European Parliaments: The First Steps*, Budapest.

Ágh, A. (1994), 'Neo-Traditionalism and Populism from Above in East Central Europe', Papers on Democratic Transitions, no. 94.

Ágh, A. (1994), 'Citizenship and Civil Society in Central Europe' in Steenbergen, B. van (ed.), *The Condition of Citizenship*, Sage, London.

Ágh, A., Szarvas, L. and Vass L. (1995), 'The Europeanization of the Hungarian Polity' in Ágh, A. and Kurtan, S. (eds.), *Democratization and Europeanization in Hungary: The First Parliament, 1990–1994*, Budapest.

Angelusz, R. and Tardos, R. (1991), 'Political and Cultural Dividing Lines among Voters for the Parliamentary Parties', *Research Review*, no. 3.

Angelusz, R. and Tardos, R (1992), 'Hitek, pártok, politikák, Ideológiai átrendezödések Magyarországon', *Világosság*, no.2.

Bruszt, L. (1989), 'Without Us but for Us?' Political Orientation in Hungary in the Period of Late Paternalism', *Research Review*, Budapest.

Cotta, M. (1991), 'Transitions to Democracy and the Building of the New Party Systems: The East European Cases in Comparative Perspective', Paper prestented at Joint Sessions of ECPR Workshops, University of Essex (22 – 28 March).

Csepeli, G., Keri, L. and Stump, I. (eds.) (1991), *How to be a Democrat in a Post-Communist Society*, Institute for Political Science, Budapest.

Hankiss, E. (1989), 'Between two Worlds', *Research Review*, no. 2.

Hankiss, E. (1990), *East European Alternatives*, Oxford University Press, Oxford.

Hankiss, E. (1990), 'In Search for a Paradigm', *Daedalus*, Winter.

Ilonszki, G. and Kurtán, S. (1992), 'Traurige Revolution – freudlose Demokratie. Aspekte der ungarischen politischen Kultur in der Periode des Systemwechsels' in Gerlich, P., Plasser, F. and Ulram, P. A. (eds.), *Regimewechsel. Demokratisierung und politische Kultur in Ost-Mitteleuropa*, Böhlau, Vienna.

Pérez-Diaz, V. (1990), 'The Emergence of Democratic Spain and the "Invention" of a Democratic Tradition', Working Papers of the Instituto Juan March de Estudios y Investigaciones, Madrid, June.

Plasser, F. and Ulram, P. A. (1992), 'Zwischen Desillusionierung und Konsolidierung' in Gerlich, P., Plasser, F. and Ulram, P. A. (eds.), *Regimewechsel. Demokratisierung und politische Kultur in Ost-Mitteleuropa*, Böhlau, Vienna.

Pridham, G. (1990), 'Southern European Democracies on the Road to Consolidation' in Pridham, G. (ed.), *Seducing Democracy. Political Parties and Democratic Consolidation in Southern Europe,* Routledge, London and New York.

Pridham, G. (1991), 'Southern European Models of Democratic Transition and Inter-regional Comparisons: A Precedent for Eastern Europe?', Paper presented at Joint Sessions of ECPR Workshops, University of Essex (22–28 March).

Rustow, D. (1970), 'Transition to Democracy', *Comparative Politics*, April, pp. 344ff.

Schmitter, P.C. and Karl, T. (1991), 'Models of Transition in South and Central America, Southern Europe and Eastern Europe', *International Social Science Journal*, no. 128, May.

Schmitter, P.C. and Karl, T. (1992), 'The Types of Democracy Emerging in Southern and Eastern Europe and South and Central America' in Volten, P. (ed.), *Bound to Change: Consolidating Democracy in East Central Europe*, Institute for East-West Studies, New York.

Schöpflin, G. (1990), 'The Political Traditions of Eastern Europe', *Daedalus,* Winter.

Appendix

Party names and acronyms

CDPP = Christian Democratic People's Party
ISP = Independent Smallholders' Party
HDF = Hungarian Democratic Forum
HSP = Hungarian Socialist Party
AFD = Alliance of Free Democrats
Fidesz = Alliance of Young Democrats

(1) PC as accumulated IN and/or external influence	(2) IN as materialization of PC's long term paradigms
field of inter actions	
(3) SM as mass reactions against dominant cultural paradigms	(4) PS as institutional indoctrination of dominant cultural paradigms

PC = Political Culture
SM = Social Movements
IN = Institutions
PS = Political Socialization
Political culture appears historically in a large variety of institutions, social movements and political socialization mechanisms.

Figure 1 History (longue durée)

(1) IN as objectivized social movements and/or forced upon from without	(2) PS as institutional programs imposed upon the population
field : of	
inter : actions	
(3) SM as organized movements ossified by institutions	(4) PC as long-term paradigms opposed to short-term socializations

Institutions shape political socialization mechanisms and mold social movements,
but are unable to change historical paradigms and continuities of political culture.

Figure 2 Political regimes (status quo systems)

(1) SM as accumulated PC and/or demonstration effects	(2) IN as re-institutionalization shaped by successful movements
field : of	
inter : actions	
(3) PS as new ruling values and socialization mechanisms	(4) PC as immune reactions of long- term cultural paradigms to IN and PS

Social movements change institutions and socialization mechanisms, but the inertia of
political culture as a set of values and paradigms prevails until the consolidation period.

Figure 3 Systemic change (or political transformations)

	compromise	violence (force)
elite (crafting)	(corporatist democracy) pact	(populist democracy) octroi
masses (mobili- zation)	reform (consociational democracy)	revolution (electoralist democracy)

Figure 4 Elite-mass dimension or types of democracy

10 The Culture of Protest. Hungarian Social Movements in Transition

Máté Szabó

The following account of Hungary's new "protest potential" during the first legislative period of its freely elected parliament (1990–1994) is based on content analyses of newspaper coverage of protest actions made by students in the Department of Political Sciences at Eötvös Loránd University, Budapest. Its purpose is to present a selective overview rather than an exhaustive analysis of new movements and forms of protest, and to initiate a broader debate on protest in post-communist democracies. This is also why references have generally been restricted to scholarly publications. I will first survey a selection of such movements, then attempt an analysis on the basis of internationally comparable criteria, and finally summarize the main characteristics of political spaces and opportunity structures for social movements in post-communist Hungarian democracy.

Social movements: a new phenomenon?

Mobilization of social movements and voices of discontent is by no means a new phenomenon in Hungary. Protests of this kind also occurred in times of crisis during the Kádár regime. In the new democracy, however, these movements and conflict potentials have been substantially transformed and restructured. In a pluralist democracy, there is no political space for a civil rights opposition and secondary or "samizdat" public sphere concomitant with a communist regime. Nevertheless, there are today civil rights movements like Amnesty International, the Raoul Wallenberg Association or the "Democratic Charta" actively countering racism and other forms of civil rights violations. In addition, former opposition groups have been institutionalized in new democratic organizations such as parties, associations and foundations.

Hungary's civic opposition of the 1970s has established itself as a party called Alliance of Free Democrats, while the "younger" generation politically active during the ensuing decade has founded the Alliance of Young Democrats. Especially this younger generation within Hungarian opposition has been supported by ecology and peace movements in its political protest, and has drawn support from the student movement for self-governed halls of residence. All these ecological, peace and student movements, however, are today less relevant for political mobilization than they were during the 1980s.

New socio-political conflicts have inevitably opened up new potentials for mobilization. The "taxi driver blockade" of October 1990, for instance, was a nationwide protest against the drastic rise in fuel prices. Though organized mainly by taxi drivers and others earning their livelihood in the transportation industry, this movement enjoyed much broader support and galvanized a public suffering from the social consequences of marketization. Taxi drivers set up blockades throughout the country, thus paralyzing the country's transport system for three days until a compromise was reached.

The Democratic Charta, an umbrella organization of various citizen's groups, has been staging public protests against anti-democratic tendencies such as racism or the attempt to impose government control on the media since 1991. It has done so with considerable success as well as support from some political parties in opposition during this period.

In 1992–3, the Society of People Beneath the Living Standard organized nationwide campaigns against the introduction of value added tax and, more generally, against "unsocial" economic policies of the government. Although this movement petered out after the general introduction of VAT, the conflict and protest potentials of socially disadvantaged groups in Hungary is still a force to be reckoned with.

"Skinheads" similarly use the political space offered by Hungary's new democracy for social and political protest, resorting to activities ranging from street fighting to supporting right wing political parties. The existence of clandestine right wing organizations with international contacts has been proved beyond doubt in a number of related trials. But political violence has not yet reached the level of organized terrorism, resembling instead German or Italian "autonomous" groups and their street riots, wilful counter-demonstrations and vandalism.

So-called alternative movements first appeared during the latter half of the 1980s and played a prominent part in the protest against the Kádár regime. A new generation of such movements has come with the democratization of Hungary, and may best be described as a network of highly interactive feminist, ecologist, anti-militarist and gay groups.

Hungary also has an abiding populist protest potential. Organizations and movements of this kind were first integrated in the frameworks of various parties, and later turned into political associations of their own. "Populism" conceived as a "Hungarian alternative" between Eastern and Western political paradigms is rooted in the country's political culture of the interwar period. It survived communist rule as a minority creed or movement of intellectuals, and has firmly reestablished itself since the advent of democracy.

The most salient feature of political protest since the Kádár regime's demise has been the institutionalization of dissent as openly staged and legal mobilization. The regime tried, with partial success, to prevent dissent from networking, gaining publicity and from establishing international contacts. Now, within the legal framework of freedom of association and public gathering, such networking, resource mobilization and open protest are widely accepted modes of public action. Beyond the framework of such institutionalized dissent, calls for civil disobedience have come from a variety of groups. Violent or illegal protest, however, has been rare in Hungary, as have clashes with the police at rallies during the last four years.

Some characteristics of protest movements in international comparison

The following internal criteria for a comparison may be established:

1. Origins and "causes" of emergence
2. Dynamics of mobilization
3. Patterns of organization
4. Forms and strategies of political action
5. Aims and programmes
6. Ideologies and utopias

Causes

In post-communist systems, a wide range of socio-political conflicts mobilize social movements and their articulation of protest. Individuals and groups face partly overlapping social, cultural and economic crises which they are unable to resolve within the framework of existing cultural, economic and political institutions and practices. The crisis of politics, or rather, the crisis of crisis management, is a permanent, ineluctable and structural feature of post-communist societies, since their political system is burdened by all types of such social, cultural and economic conflicts. Consequently, the political space for spontaneous, unorganized and innovative forms of expression, for assertive and vociferous social movements, is widening continuously.

Within this general setting, a number of situative factors have helped to shape Hungarian movements and their specificities discussed here. The rigidly bureaucratic and authoritarian character and political style of Christian nationalist government in Hungary between 1990 and 1994 drew persistent criticism from the Democratic Charta and other alternative movements. The social costs and consequences of economic restructuration prompted, as mentioned above, LAET, the Society of People Beneath the Living Standard to rally support for its members' cause. Youth unemployment, the general crisis of values, and problems of political culture led to the development of skinhead radicalism. Parts of the middle classes also proved susceptible to this pull in the face of the hardships of transformation and a rehabilitation slow in coming. A new nationalist middle class found its concerns echoed in the language of nationalist radicals or populists, most notably on the occasion of controversial legislation concerning government control of the media in 1991–1992.

According to Rammstedt (1979), social movements try to secure unity of aims and causes. They "solve" conflicts and crises by recourse to their own programmes. Movements, and especially protest movements tend to articulate complex socio-political problems in sets of "yes or no" questions, in simple dilemmas or alternatives, and try to rally support for a political activism attacking a problem "at its roots".

Dynamics of mobilization

Social movements should not be confused with mass rallies or demonstrations, even though they often resort to this type of public action in pursuit of their goals. Mass mobilization *per se* is bound to a particular situation, and it represents a spontaneous and transient form of collective behaviour. Social movements, in contrast, are and transient form of collective behaviour. Social movements, in contrast, are characterised by stability, persistence, and by institutionalization of aims, organization and strategies. The "resource mobilization approach" of social movements concentrates on networks, communities, and similar forms of collective initiative preceding the level of massive protest rallies.

These assumptions are borne out by Hungary's new social movements. Behind the Democratic Charta, there was a "network of networks" of different political groupings ranging from citizens' initiatives to parties and helping to mobilize resources for their umbrella organization. LAET's setup resembles the pattern of a "transitory team", a model applied to the analysis of social movements in the United States (Zald-McCarthy 1987, 257–8). In "transitionary teams", a small number of professional activists canvass a large circle of members and sympathizers in support of a single issue campaign. LAET consisted of a core of committed activists holding together a nationwide network. "Populism" built an organizational network of mobilization within a large political party of loose integration. The networks of skinheads and alternative groups are in turn mainly rooted in urban youth subculture and gangs.

All present-day social movements in Hungary are based on such preexisting networks or organized forms of collective action. The dynamics of mobilization is generated within formalized or informal groupings or communities, the permanent goal being the drawing together of personal and other resources in the service of particular aims.

Forms of action

In social movements, political protest is one of the few common features shared by groups otherwise pursuing an array of different strategies. Protest as a resource is a major component of any movement strategy, but protest cultures vary substantially in internal factors such as aims, tactics, organizational setup and social bases, and even more in the external factors shaping them.

Joachim Raschke (1985) has tried to depict the relationship between strategies and forms of collective action as follows:

Table 1 Strategies and forms of collective action

strategies/forms of action	intermediary action	demonstrative action	direct-coercive action
institutional	x	x	–
multidimensional	x	x	x
anti-institutional	–	x	x

According to this configuration, forms of action mediating between incumbents and movements (bargaining, pressuring, presenting signed petitions), actions consisting in spectacular publicity for the aims and claims of the movement (demonstration rallies, hunger strikes), and direct coercive forms of action (blockades, ostensive destruction or

152

vandalism, terrorism) can all be related to corresponding movement strategies. Thus, institutional strategies are based on intermediary and demonstrative types of action, but do not use coercion. Anti-institutional movements deploy demonstrative and directly coercive action strategies, but will not usually aproach incumbents with an intermediate strategy. In most cases, of course, any combination of these – institutional and anti-institutional – ideal types will be found combining all types of action in multi-dimensional strategies. Not all forms, however, fit into Raschke's scheme. Civil disobedience, for example, combines elements of all three types of action as well as of the two opposed sets of strategies.

We can nevertheless apply this scheme for a succinct analysis of action strategies adopted by Hungarian post-communist social movements. The Democratic Charta's strategy is institutional; it uses demonstrative and intermediary forms of action. LAET developed a multidimensional strategy, but more recently radicalized its range of actions to include petitions, hunger strikes, and an aggressive campaign to delegitimize and topple the constitutionally elected government. While the main protagonists of the movement were not in favour of directly coercive action, radicals within its ranks openly discussed it. The protest organization collapsed when the possibilities of institutional protest had been exhausted and radical tendencies correspondingly gained momentum (Radicals "punished" the former leader's "betrayal" by devastating his apartment). Similarly, alternative movements have mainly employed institutional strategies but have, as in the case of Earth Day 1993 in Budapest or of similar conflicts over ecological issues, resorted to coercive action such as public rallies and blockades prohibited by local authorities. Among skinheads, there is no discernibly homogeneous strategy. Some groups commit acts of political violence against minorities, cause disturbances at demonstration rallies or official ceremonies by staging counter-demonstrations or street fights. Their activities usually involve some violation of the law, so their strategy clearly has an anti-institutional thrust and direct coercive action may characterize them to some degree. Other parts of the movement, however, remain within the bounds of law and accepted forms of politics, staging legal demonstration rallies and emphasizing their institutionalism. Anti-institutional and institutional action are thus both found in the movement referred to as skinheads. Radical populists, likewise, combine both strategies. Directly coercive action such as trade on an illegal "people's market" prohibited by district authorities in Budapest 1993, or appeals to media sabotage or blockades in 1992, are examples of the use of both types of strategies by these groups.

Legal public protest as such is, of course, a new phenomenon in Hungary. It is also no great surprise that the limits of the law concerning public gathering and the forming of associations are sometimes stretched, although demonstration rallies, along with other forms public dissent, were institutionalized in 1989. There are direct-coercive forms of action used by some movements as part of a set of anti-institutional strategies, but compared with France or Germany, for instance, political violence and the abuse of freedom of speech are still marginal phenomena in Hungary. Here, too, while direct coercive action is on the rise, it is still incomparably lower than in some Western democracies.

This fact may be explained by traditions of authoritarian rule, by the strong and broad basis of legitimacy enjoyed by the new political institutions, and by a consistent attempt on the part of the authorities to avoid provoking conflicts. My research on cases of law infringement during demonstration rallies and public gatherings has shown that no cases ended up in court, despite the fact that they often featured prominently in the media. Hungarian authorities, and the police in particular, have tried to avoid carrying such conflicts into the legal sphere because of their well-earned reputation of violently attacking illegal demonstrators under communism. During the last four years, their strategy has been to avoid confrontation whenever possible and not to produce "cases" involving demonstrators at all.

Organization

The relations between movements and political organizations, as well as between dynamics of social movement organizations (SMO) vary with the groups involved. "Movement versus organization" dilemmas are patterned differently from one movement to another, as are their SMOs.

The Democratic Charta is an umbrella covering a variety of more or less organized networks. Some political organizations support this umbrella by supplying resources and infrastructure; between 1990 and 1994, opposition parties represented in parliament used some of their funds for this purpose. But this support was granted on condition that the Charta did not develop into an autonomous actor participating in electoral politics. Against a tendency within the Charta to build up a political organization and profile, established parties thus succeeded in limiting it to a citizens' initiative and guardian of human rights with no political profile or potential electorate of its own. Organizational development ist mainly concentrated on staging public rallies and periodic protest campaigns such as those against authoritarian trends, or on "civic" public holidays like 15 March commemorating the Hungarian revolution of 1848. Meanwhile, the Charta's SMO does not follow a pattern of regular and continuous activity. And since the buildup of such activities depends entirely upon the supporting organizations and groups, it has retained the character of a transient initiative lacking professional leaders and a bureaucracy, and playing the part of a coordination committee resembling those of Western peace movements during the early 1980s.

LAET based its protest rallies on an informal network of local organizers and one central but transient team taking care of national coordination. SMO development within LAET looks very much like a textbook model: activities getting under way in a community near Budapest extended centrifugally until they covered the entire nation. Intensive protest and the strategic dilemma of organization led to differentiation within the "transient" team, ending in secession of groups supporting the idea of a "poor people's party", and leaving the remainder to dissolve or retreat to local grass root milieus, especially after the failure of the anti-VAT campaign.

Especially in the case of LAET, the media in part functioned as "organizational supplements", as they had in the case of activities of US movements (Molotch, Oberschall) during the 1960s. But although media coverage of public protests does support unorganized dissent, it can easily lead movements to the verge of breakdown. Once the news value of a protest action has been exhausted, attention quickly shifts to other issues and the lack of coordination and mobilization within the movement makes itself felt. LAET was to some extent "made" by media reporting on organized hunger strikes or subscription campaigns to dissolve parliament. After the introduction of VAT, however, LAET fell into oblivion and virtually disappeared from the national public sphere.

Skinhead and alternative movements ar both grounded in loose networks of student, youth group, or other subcultural milieus. Such movements already existed in the past decade, and their dynamics is in part related to similar Western European cultural and social trends. In both cases, inistitutions and networks of urban "counter-culture", such as music, fashion, entertainment, debating societies and other milieus or media of expression are the principal modes of integration. These integrative factors can substitute missing or defunct organizational infrastructures in the course of a mobilization campaign.

Both movements built up their organizational structures after democratization. Violent skinhead subcultures had and have, as published material relating to criminal proceedings against them has shown, clandestine networks linked up with Hungarian and Western European neo-fascist underground groups. Some of the formerly clandestine organizations are now trying to legalize their activities by either joining existing right wing parties as their "youth sections" or by setting up their own formal associations, a trend clearly evident from various experiments with organization and protests of right wing groups trying to reestablish Hungarism – that is, Hungarian interwar fascism – in 1993-4. Although there are restrictions on the public display of fascist symbols in Hungary, the government's mixed reaction to violent riots and lenient law enforcement have helped skinheads' attempts to gain a measure of legitimacy. This is true despite the fact that criminal proceedings have in some cases been instigated by the state attorney. Some intermediary figures, moreover, are accepted by the media and political organizations as representatives articulating the demands arising within right wing youth subcultures. As a result of this indirect legitimation through public debate and political tolerance, we may well see a rise of organized right wing youth activism within the near future. Skinheads routinely follow demonstrating Charta members, "alternatives" or Romanies and Sinti as shadows threatening violent intervention, and their demands are publicly articulated. The menacing presence of right wing activists has not yet produced violent clashes on a massive scale. But during public ceremonies and rallies around 15 March 1994, rightist radicals staged an attack on Charta members, this being, so far as I am aware, the first violent clash between protest groups in the new democracy.

Alternative movements have also established their own formalized networks. Thus, Alternative Network was registered as an association in 1992. It is rooted in dozens of initiatives and includes anti-militarists, gay, feminist, peace, ecology and anarchist groups. AN may best be described as a coordinating board governed by principles of grass root democratic decision making. It provides member organizations with information and infrastructure, and also tries to raise funds for them. AN does not, as such, mastermind political activities. Rather, groups of activists use the network for their impromptu mobilizing and joining of scattered protest potentials. In addition, feminists, antimilitarists, and ecologists have established separate networks for specific coordinative purposes. Any campaign or protest action originating in these milieus may thus be seen as the joint effort of a number of networks, or of a semi-formalized, semi-organized network of networks.

Hungarian "populists" have gathered mainly in the Hungarian Democratic Forum (HDF), the strongest governing party in parliament between 1990 and 1994. "Radical nationalists" among them acquired a distinct profile from the very beginnings of the party, and set themselves apart from the elite which József Antall, party chairman and Hungarian prime minister until his death in 1993, had done much to integrate into its fabric. Antall's liberal and conservative policies drew strong criticism from these radicals, who also took to the streets in order to fend off opponents of the ruling Christian-nationalist coalition as diverse as blockading taxi drivers and liberally minded media executives. Through their series of demonstration rallies, their heated debates and rhetorical skirmishes with liberals, socialists and Democratic Charta representatives, populists effectively turned into a "movement within the organization" (Zald-McCarthy 1987, 185–223) of HDF.

The growing distance between populists and professional politicians holding seats in parliament and positions in government eventually led to an organizational split running through HDF. A protracted conflict between leaders and the populist faction resulted in disciplinary measures against the latter, who insisted on autonomous organization and policy stances. Populists set up their own Movement for the Hungarian Alternative (Magyar Ut Mozgalom), which comprised a network of foundations and initiatives. After the decisive clash with HDF's leadership, they finally established a Party of Hungarian Life and Truth (Magyar Igazság és Elet Pártija, MIEP).

Radical populism retained its character as a protest movement throughout the time it formed part of HDF as a "movement within organization". Its roots go back to the interwar period, and its traditions survived communism mainly as a movement of intellectuals. Repoliticization of this literary movement coincided with democratization, and several groups at once competed for Hungary's "true" post-communist populist heritage. Populists never quite lost their unease with party structures, and maintained loose cultural movements and groups alongside frameworks of party or other types of political organization. As a consequence, the populist milieu offers a wealth of cross-cutting and overlapping political, social, and cultural associations.

Such a diversity of gathering patterns is the rule in Hungary's post-communist political movements. In addition to the examples given, the dilemmas of "party versus movement" can be found in almost all of todays parties in and outside parliament. There is a lively interplay between parties and various social movements or their SMOs in Hungary's political system, and similar interaction has developed on the level of trade unions, pressure groups and organized interests. Social movements are an important factor in Hungarian politics today. They are worthy subjects for students of political organization in post-communist democracies, especially since important political parties, trade unions, pressure groups and organized interests are all rooted in such movements.

Aims and claims

Hungarian social movements pursue a wide range of aims. The Democratic Charta, for instance, has a clearly articulated value system in the form of a "charta", and through its representatives or coordinating committees defines issues accordingly. Isolated, as well as serialized, protest actions have thus targeted racism, infringements on freedom of speech, biased reporting in public service media, violations of minority rights, etc. The Charta's main orientation may be defined as the defence of democratic rights and values against political trends endangering them. Issues of world politics, of war and peace, or of socio-economic transformation were therefore not high up on the Charta's agenda.

LAET had no general charta of values, but instead devoted itself to a single issue, namely to criticism of value added tax. It proved incapable of addressing new issues despite the fact that the social consequences of marketization offered a wide range of them.

The Charta and LAET differ significantly in both their aims and their strategies. While the former targets a variety of issues, its strategic repertoire is limited to a repeating pattern of subscription campaigns and public rallies. The latter, in contrast, focused on a single issue, but went through stages of radicalization in its strategies until final defeat and disorientation. The Charta has become an institutionalized actor of protest culture within a few years, while LAET's life span ended after a few months of its only campaign. Although the Charta's aims are primarily situated in the sphere of politics or political culture, its founding document also recognizes the importance of some social issues. Conversely, LAET began as a socio-economic protest organization, became politicized through its own dynamics, and ended up rejecting the political system of representative democracy entirely. Whatever their differences, however, both movements are campaign- and action centered, rather than being grounded in a networks of local and community organization.

In contrast, alternative and skinhead movements articulate socio-cultural rather than political or socio-economic claims. Their actions and demands take shape in the context of subcultural milieus, both in the case of nonviolent, antimilitarists and of violent, paramilitary youth gangs. Their first concern is not with politics, but with the creation of life styles and subcultural value systems. Political or socio-economic questions are usually addressed by them only insofar as they relate to these concerns. The kind of politics which concerns them has been instrumentalized mainly by political organizations outside the subcultural movement. Since they are unable to play their own part in power politics, it is either the Charta and small liberal or ecology groups, or right wing organizations, which make politics with and for them.

Populism, as mentioned above, strives for continuity with its venerable Hungarian ancestry, and derives its goals and demands from that tradition. While a "movement within organization", populists claimed to be the true nationalist radicals, and consistently criticized HDF's manifesto and government policies. They articulated anti-liberal and anti-communist attitudes as well as the idea of an Hungarian alternative or "third road" between capitalism and socialism. When populist fundamentalism turned into a movement and party of its own, its criticisms of HDF and of all liberal and leftist positions were amalgamated into a rhetoric of rejection: the communist past and the four years of Hungarian democracy are coming to resemble each other in their alleged betrayal of truly Hungarian principles and disappointment of the "masses", at least when viewed through the populist prism. A "real" system transformation is advocated as a populist solution of political, social, and economic problems alike. Not unlike skinheads, populists hold strong nationalist and racist views and attack Zionism as well as "Jewish rule" in Hungarian society and politics. Of all movements discussed here, populism is the most fervently ideological.

Ideologies and utopias

In the political thought and articulation of social movements, ideologies and utopian visions immediately define and shape goals and demands. They all espouse strong views of "the good life" in the form of general values or explicit models of alternative socio-political systems. Correspondingly, they either express wholesale rejection of the "existing order", or focus on specific aspects of the political system with a view to adjustments in its institutions. More often than not, their criteria for sorting the good and the bad in politics are taken from old traditions in the history of political and social thought, and are inscribed in the historical unfolding of certain such "ideologies and utopias". Movements sometimes invoke historical predecessors or align themselves with contemporary brothers and their thought and action patterns. Historical, international or regional intellectual and social currents in any combination may be taken up and integrated into their discourse. Textually speaking, committed intellectuals produce coherent ideologies and utopias for various movements, not all of whom, however, can or wish to rely on such intellectuals and such utopias. Many socio-political initiatives employ symbols, slogans, and lifestyles to state their claims internally and externally.

In Hungarian movements generally, few explicitly formulated ideologies and utopias can be found. Rather, their visions of the good life and the main thrust of their criticism of political reality are implied in declarations, manifestos, intellectual profiles of leading figures, forms of action, strategies of debate, and in references to historical or international examples. As such, they may be reconstructed from a movement's discourse, and then set in the context of traditions and current trends in political and social thought.

In the case of the Democratic Charta, the leading figures and personalities are well-known Hungarian intellectuals whose works exemplify a liberal and democratic miminal consensus that may be identified as the essence of "Chartist" mentality. Any racist, nationalist, populist or other movement posing a threat to values of civility, liberality or democracy becomes a target of Charta criticism. In addition, Chartists have spoken out against "Hungarian backwardness", glorification of pre-communist Hungarian political culture, and traditions of authoritarianism. Its intellectual ancestry is to be found in 19th century Hungarian democratic radicalism, as well as in civil rights dissidence under communism. Internationally, civic movements against racism and for human rights, such as Amnesty International and various local and national initiatives, bear close resemblance to the Charta and have served the latter as blueprints. Documents of of liberal-democratic opposition and political emigration during the communist period have lately been (re)published and widely distributed throughout Hungary.

LAET could not enlist intellectuals in its effort. It was a movement set up by socially disadvantaged and politically marginalized local activists, who even rejected the influence and domination of literary urbanites. But the declarations of its leaders and the character of its protest actions mirrors concepts of "popular justice", "primitive rebellion", and "people's democracy" independent from, and markedly different to, communist taditions. The idea of a "moral economy" is inherent in the very naming of the movement as an "association of people beneath the living standard". Besides rejection of particular government policies, such as the introduction of value added tax, LAET articulated a fundamental disapproval with Hungarian governance. In its attempt to initiate a referendum aiming at the dissolution of parliament as an institution, LAET challenged the legitimacy of parliament and representative democracy alike. Its campaign was devoted to the establishment of "radical democracy" and concomitant institutions of "popular sovereignty"; and it presented itself as a movement of "real" people, as against the corrupt, selfish political elite and its institutions. Such ideas can be traced to earlier populisms of various persuasions, and to "poor peoples' movements" of Western democracies. So shortly after the demise of communism, of course, any reference to communist or socialist traditions was rejected by LAET.

Hungarian Skinheads see themselves as part of an international trend among Western youths and try to adapt their values and means to an Hungarian context. One of their main enemy concepts and targets has been Hungary's Romany and Sinti population, as well as travelling coloured people living in urban centers and with no permanent residence. A long tradition of antisemitism finds continuity in the skinhead movement. Anti-communism is also strongly articulated. In their search for positive value concepts, skinheads revert to "pure" Hungarian traditions of nationalism, militarism, irredentism, and "Hungarism" or Hungarian fascism in combination with other, mainly German, ideas. There are no intellectuals within their ranks, but skinhead music and fashion are common currency in urban youth subcultures, and their political discourses tie in with those of other political right wing milieus.

Hungary's alternative movements have in the main copied international trends in their choice of positive and negative social or political orientations. Intellectuals, especially young urban ones, are somewhat overrepresented among the supporters and activists of such movements. As a consequence, "alternative" issues are widely discussed and disseminated in the political, cultural and academic spheres. Intermediate and well-known intellectuals are as important in popularizing alternative values and concepts as they are in the West. Despite resting on principles of grassroot democracy and non-violent resistance, alternative movements do come in conflict with institutions of democracy and the rule of law. In contrast to the situation in Western democracies, Hungarian alternative movements usually support demands made by the Democratic Charta. That is, the slogan of "civil society" advanced by democratic opponents and alternative movements in their struggle against communist rule remains a common point of reference for two types of movements which are not usually found cooperating in Western democracies. Liberal radicalism – in American parlance – and civic movements resemble post-communist alternative subcultures closely in some of their political stances. They share a criticism of *etatisme*, bureaucracy, authoritarian tendencies, militarism, and racism, which centre on the idea of civil society. Liberal civic and anarchic alternative movements joined forces and let their differences rest, quite in contrast to comparable movements and their conflicts in Western democracies. Human rights, tolerance, and civic values are their overriding concerns and common denominators. In the sphere of organized party politics, however, relations between liberals and alternatives are somewhat more strained. Within the Alliance of Young Democrats – a party established by protest movements of the 1980 – liberals have consistently marginalized alternatives.

Hungarian populism, as mentioned above, is today a strongly and consciously ideological movement, which may be the result of its "catacomb existence" as a subdued cultural movement under communism. Interwar populism maintained relations with fascist and communist groups alike, and was somewhat torn between, respectively, their racism and social radicalism. Some populist intellectuals could therefore easily be integrated into communism after their own polical organizations had been suppressed. The new populism has been set up by intellectuals, who are mainly concerned with continuing interwar traditions. Although interaction with populisms abroad are rare, the movement can be compared with other, so-called *narodnik* currents and their nationalist, traditionalist, racist, anti-Western value systems encountered throughout Eastern Europe and Russia in particular. "Hungarian alternative" as a positive orientation, and intellectual, economic, or cultural domination by liberals and leftists suspected of "foreign" (Russian, German, Jewish) origin as a negative projection largely define the populist mindset, as does the craving for a "real revolution" rolling back advances made by its numerous enemies. Much of the writing in pamphlets, political publications, and literature illustrates Hungarian populism's ideologies and utopias of the years between 1990 and 1994. In a wave of revivalism, some classics of populism in emigration have recently been made available in reprints.

Hungarian intellectual historians are currently debating the hypothesis that the country has never had a "Hungarian ideology and utopia". Its social an intellectual movements, it is said, have all been "pragmatic" and as such lacked a philosophy of their own, adopting and combining instead Eastern and Western currents of thought. According to criteria of originality, nationalism and populism are the only candidates with a possible claim to the status of "Hungarian ideology". At the same time, the main modern ideological currents and strands have accompanied Hungarian socio-political development since 1798, and there are interesting cross-relations between these currents and the manifestos and discourses of social movements. Since 1989, this has again been obvious in social movements.

Social movements thus tend to undergird their approach to issues with utopian visions, a practice which sometimes also leads to dilemmas and incompatibilities between "teleological" general beliefs and particular demands.

As mentioned above, mobilization of social movements also occurred under the Kádár regime, but the latter's opportunity structure meant that it had to happen in a sphere of illegality, and with no access to public spheres and large audiences. As a shadow publicity, it could only have marginal effect until the advent of constitutionalism, rule of law, free and fair elections, and the right to associate and assemble publicly. Between 1990 and 1994, conditions for mobilization have, in the main, matched those of Western democracies. By acting publicly and in order to fit into the new framework of representative democracy, social movements have had to undergo a number of transformations in their actions, patterns of mobilization, and in the shape of their political criticism. I will end with a brief outline of both the nature of the changes and the consequences they have had.

Legalization

Under the Kádár regime, all types of autonomous socio-political mobilizations were forced underground, since there was simply no legal framework for associations, trade unions, or political parties. Now, any movement has a choice of several organizational forms laid down constitutionally and anchored in the political system. While any non-conformist initiative prior to 1989 had to resort to informal, clandestine networking, the decision whether to form a movement or an organization has become a major stategic dilemma for former opposition movements. The new political opportunity structure offers several paths to institutionalization, and becomes a challenge for many movements by doing so.

Differentiation

Anti-communist opposition movements were forced into solidarity with their kind. Their common enemy was the party state supervising and suppressing them, and the burden of restriction produced a sense of "brotherhood" among them in spite of their widely differing values and aspirations. With democratization, the common enemy disappeared and the publicly articulated conflicts between movements began to break up that solidarity. Mobilization is now widespread and diverse, and movements now compete in, for the same goods of support and resources in the same political market. Indeed, they have become opponents or even enemies for good reasons.

Networking

Cooperation of movements, as well as the sharing of resources and infrastructures, was systematically hampered under the old regime in order to stave off a multiplication of protest potentials. Hungarian movements could expect a certain degree of lenience on the part of the authorities in comparison with, say, the GDR or Romania. Networking, however, was suppressed immediately wherever the opportunity arose. What cooperation there was despite such repression could only take place publicly after democratization. As a new phenomenon, networking is spreading rapidly as a means to circumvent scarcity of resources and to capitalize on synergies. Again, becoming formally established and "going public" is a process generating conflict and necessitating difficult adjustments. But it has also made networks more flexible and open to alliances or federations. In some recent campaigns, for example, networks of networks have been established for the purpose of a single campaign.

Internationalization

International cooperation between movements was, of course, the primary target of communist repression. Despite the internationalism of working class movements, communist governments effectively prevented resource mobilization or exchange of ideas on an international level. State control of communication flows set up barriers against global as well as regional cultural and political trends. Now groupings within civil society have unrestricted access to international networks, which are proving particularly important for new movements lacking resources and know-how. "Poor" Eastern movements are now looking for more well endowed Western partners, but also · elsewhere, in their attempts to augment resources. As former opposition movements, they are thus having to adapt to the breakup of small and tightly knit circles. In particular, such relations have been established with Hungarian exiles in Western countries, as well as with members of Hungarian minorities in neighbouring countries.

Legality and legitimacy of protest

In post-communist democracies, a new conflict between legality and legitimacy has arisen. Under communism, of course, giving political directions was a monopoly affair. All non-official forms of public expression were illegal by definition. But they were far from illegitmate in the eyes of as public alienated from communist power structures and in part highly responsive to democratic values enshrined in, say, UN declarations or the Helsinki Charta. But since legal space for political dissent has now been opened up, new ways of legitimizing protest strategies surpassing the legal framework must be found. "Civil society" and "peaceful resistance" circulate among intellectuals as possible sources of such legitimacy. But the question of whether and how to grant legitimacy to illegal activity is not an academic one. Even a quick look at the widening repertoire of protest strategies breaking the law will show that they profoundly challenge public and constitutional order. Among illegal forms of protest, violence is not the most salient, and has in fact remained a marginal phenomenon involving mostly clashes between domonstrators and the police. Political dissent, once "entirely outlawed but mostly legitimate", has now become "mostly legal, sometimes illegal, but rarely legitimate" – a tremendous shift in the relationship between actors and their public. Concepts and criteria of legitimacy in political protest therefore need to be worked out quickly, and a new political culture of dissent building on traditions of underground opposition, but able to deal with new conflicts and establish new forms of consensus, must be found.

To adapt Albert O. Hirschman's famous "exit" and "voice" options for behaviour in crisis-ridden organizations: opposition under communism had no choice but to exit. Now, the decision between, or particular combination of, exit and voice strategies must be a conscious one. New opportunity structures also challenge the self image and identity of protest movements. As new conflicts arise in the context of new social, economic, and political conditions, new modes of conflict will be needed to articulate them.

Conclusions

We have seen that since 1990, Hungary has had a complex and differentiated sector of socio-political movement operating within the political system. These movements make widely differing and partly conflicting claims, and are prominent voices in current Hungarian political discourse. The "culture of protest", as a functional segment of political culture is still in the making, and there are a number of obstacles which will be overcome only in a slow process of political apprenticeship. Political and administrative authorities are still partly excluded from the "consent of democrats", a demarcation transcending that between government and parliamentary opposition. Some movements, moreover, use violence (as in the case of skinheads) or privileged access to governing elites (radicals within HDF before their secession) to appropriate political space. Although the political and administrative control of social movements has been, should be, and will be pluralized as well as democratized, their protagonists' decisive socialization experiences all date back to the times of authoritarianism. But the process of political learning in Hungary is well under way, with a pluralist sector of social movements functioning as one of its main agents.

References

Bayer, J. and Deppe, R. (eds.) (1993), *Der Schock der Freiheit. Ungarn auf dem Weg in die Demokratie*, Suhrkamp, Frankfurt.

Bilecz, E. (1987), 'A magyart társadalmi mozgalmak fejlödéséröl', *Ifjúsági Szemle*, no. 6, pp. 22–33.

Bozóki, A. (1988), 'Critical Movements and Ideologies in Hungary', *Südosteuropa*, nos. 7–8, pp. 377–388.

Bozóki, A. (1992), 'Democrats Against Democracy. Civil Protest in Hungary since 1990', in Szoboszlai, G. (ed.), *Flying Blind*, HPSA, Budapest, pp. 382–397.

Bozóki, A. (1992a), 'Demokraten gegen Demokratie? Ziviler Protest in Ungarn seit 1990', *Berliner Debatte*, no. 5, pp. 60–70.

Bozóki, A. (ed.) (1992b), Tiszta lappal. a FIDESZ a magyar politikában, 1988–1991, FIDESZ, Budapest.

Bozóki, A. et. al. (eds.)(1992c), *Post-Communist Transition. Emerging Pluralism in Hungary*, Pinter, London and New York.

Bozóki, A. (1993), 'Az Anarchista Ujság', *Mozgó Világ*, no. 12., pp. 129–141.

Brand, K.-W. (1990), 'Massendemokratischer Aufbruch im Osten: Eine Herausforderung für die NSB-Forschung', *Forschungsjournal Neue Soziale Bewegungen* , vol. 3, no. 2, pp. 9–17.

Brunner, G. (1993), *Ungarn auf dem Weg der Demokratie*, Bouvier Verlag, Bonn.

Dalos, G. (1986), 'Die kurze Geschichte der ungarischen Friedensgruppe "Dialog"', *Perspektiven des Demokratischen Sozialismus*, no. 3, pp. 187–197.

Ehring, K. and Hücking, H. (1983), 'Die neue Friedensbewegung in Ungarn', in Steinweg, R. (ed.), *Faszination der Gewalt. Friedensanalysen 17*, Suhrkamp, Frankfurt, pp. 313–350.

Fehr, H. (1993), 'Soziale Bewegungen im Übergang zu politischen Parteien in Ost-Mitteleuropa', *Forschungsjournal Neue Soziale Bewegungen,* vol.6, no. 2, pp. 25–41.

Fisher, D. and Davis, C. (eds.) (1992), 'Civil Society and the Environment in Central and Eastern Europe', *Ecological Studies Institute*, London, pp. 51–97.

Fleischer, T. (1992), 'A dunai vízlépcső esete', *Társadalomkutatás*, nos. 2–3, pp. 28–48.

Frentzel-Zagorskq, J. (1990), 'Civil Society in Poland and Hungary', *Soviet Studies,* vol. 42, no. 4, pp. 759–777.

Haraszti, M. (1990), 'The Beginning of Civil Society: The Independent Peace Movement and the Danube Movement in Hungary', in Tismaneanu, Vladimir (ed.), *In Search of Civil Society*. Routledge, Chapman Hall, New York and London, pp. 71–88.

Hirschmann, A. O. (1970), *Exit, Voice, Loyalty*, Harvard University Press, Cambridge, MA.

Hirschmann, A. O. (1984), *Engagement und Enttäuschung*. Suhrkamp, Frankfurt.

Juhász, J. (1992), 'Ökologische Konflikte in der Transformation Ungarns', in Nissen (1992), pp. 163–168.

Kende, P. and Smolar, A., 'Die Rolle oppositioneller Gruppen am Vorabend der Demokratisierung in Polen und Ungarn (1987–1989)', *Forschungsprojekt Krisen in den Systemen sowjetischen Typs*, nos.17–19, INDEX, Köln.

Kitschelt, H. P. (1986),'Political Opportunity Structures and Political Protest', *British Journal of Political Science*, vol. 16, no. 1.

Knabe, H. (1988), 'Neue soziale Bewegungen im Sozialismus', *Kölner Zeitschrift für Soziologie und Sozialpsychologie*, no. 3.

Knabe, H. (ed.) (1990), 'Soziale Bewegungen und politischer Wandel im Osten', *Forschungsjournal Neue Soziale Bewegungen*, no. 2, special issue.

Machos, C. (1992), 'Von den "alten" zur "neuen" ungarischen Opposition', Demokratische Charta 91, *Berliner Debatte*, no. 4, pp. 57–68.

Machos, C. (1993), 'FIDESZ – Der Bund Junger Demokraten. Zum Portrait einer Generationspartei', *Südosteuropa*, no. 1, pp. 1–26.

Miszlivetz, F. (1989), 'Emerging Grassroots Movements in Eastern Europe: Toward a Civil Society?' in Gáthy, Vera (ed.), *State and Civil Society*, MTA, Budapest, pp. 99–113.

Nissen, S. (ed.) (1992), *Modernisierung nach dem Sozialismus*, Metropolis, Marburg, pp. 141–163.

Nugent, M. L. (1992), *From Leninism to Freedom*, Westview Press, Boulder.

Paetzke, H.-H. (1986), *Andersdenkende in Ungarn*, Suhrkamp, Frankfurt.

Pakulski, J. (1991), *Social Movements*, Longman Cheshire, Melbourne.

Pross, H. (1992), *Protestgesellschaft*, Artemis Winkler, München.

Rácz, M. (1993), 'A magyar gerillasajtóról', *Mozgó Világ*, no. 12, pp. 141–147.

Ramet, S. P. (1991), *Social Currents in Eastern Europe*, Duke University Press, Durham and London.

Rammstedt, O. (1979), *Soziale Bewegung*, Suhrkamp, Frankfurt.

Raschke, J. (1985), *Soziale Bewegungen*, Campus, Frankfurt.

Rucht, D. (ed.) (1991), *Research on Social Movements*, Campus, Frankfurt.

Rucht, D. (1991a.), 'Das Kräftefeld soziale Bewegungen, Gegenbewegungen und Staat', *Forschungsjournal neue Soziale Bewegungen*, no.2, pp. 9–17.

Schöpflin, G. (1979), 'Opposition and Para-Opposition; Critical Currents in Hungary,1968–1978', in Tökes, R. (ed.) (1979), *Opposition in Eastern Europe*, MacMillan, London, pp. 142–187.

Sólyom, L. (1988), 'Hungary: Citizens' Participation in the Environmental Movement', *IFDA-Dossier*, no. 64, pp. 23–35.

Stumpf, I. (1988), 'A szakkollágiumokról', *Ifúsági Szemle*, no. 1, pp. 54–69.

Stumpf, I. (1992), 'Youth and Politics' in: Gazsó, F. and Stumpf, I. (eds.), *Youth and the Change of Regime*, Institute for Political Sciences, Budapest, pp. 27–37.

Szabó, M. (1991a.), 'Changing Patterns within the Mobilization of Alternative Movements in Hungary' in Szoboszlai, G. (ed.), *Transformations to Democracy*, HPSA, Budapest.

Szabó, M. (1991), 'Die Rolle von sozialen Bewegungen im Systemwandel in Osteuropa: ein Vergleich zwischen Ungarn, Polen und der DDR', *Österreichische Zeitschrift für Politikwissenschaft*, vol. 20, no. 3, pp. 275–289.

Szabó, M. (1992), 'The Taxi Driver Demonstration in Hungary: Social Protest and Policy Change' in Szoboszlai, György (ed.), *Flying Blind*, HPSA, Budapest.

Szirmai, V. (1992), 'Die Rolle ökologisch-sozialer Bewegungen in Ungarn' in Nissen 1992, 141–163.

Tamás G. M. (1993), 'A demokratikus ellenzék hagyatéka', *Világosság*, nos. 8–9, pp. 137–151.

Tismaneanu, V. (ed.) (1990), *In Search of Civil Society*, Routledge, New York.

Tökés, R. L.(ed.) (1979), *Opposition in Eastern Europe*, MacMillan, London, pp. 60–113.

Varga, J. (1986), 'Warum kämpfen wir gegen das geplante Donau-Kraftwerk von Nagymaros?', *Perspektiven des Demokratischen Sozialismus*, no. 3, pp. 197–205.

Zald N. and Mayer-McCarthy, J. (1987),. *Social Movements in an Organizational Society*, Transaction, New Brunswick.

Zimmermann, E. (1983), *Massenmobilisierung. Protest als politische Gewalt*, Edition Interform, Zürich.

SLOVAKIA

11 Understanding Slovak Political Culture

Silvia Mihalikova

Recently, students of political culture and other commentators have been differentiating post-communist European states on the basis of national characteristics of democratization. During his visit to Slovakia in July 1993, for instance, Z. Brzezinski opined that the country did not fulfill political and economic criteria for continuous democratization, and was to be ranked among unstable Balkan countries such as Albania or Romania. In his 1994 Crans Motana speech, Czech prime minister Vaclav Klaus similarly expressed doubts concerning Slovakia's progress towards democracy and market economy. Politicians or political scientists from traditional democracies often attempt to forge blueprints for countries with inferior or fragile democratic institutions or practices to follow, or at least offer evaluative criteria for judging the quality of a given democracy. Inevitably, such approaches raise questions as to the content and legitimacy of value concepts like "quality of democracy". Who decides on democracy's state of the art, and on what grounds can countries and their political systems be assigned a place in this new hierarchy?

Slovakia is commonly referred to as both a case of peaceful management of the breakup of a country, and as one of a culmination of typical post-communist transition problems: nationalism, xenophobia and revived authoritarianism. After four years of post-communist transition and about a year of independent statehood, Slovaks' responses to the changes following the collapse of communism in 1989 in the main correspond to those of other ECE countries. In part, however, they differ markedly as a result of the country's specific history and political situation. There is widespread uncertainty with respect to the level or state of democratization within Slovak society; some commmentators, as we have seen, altogether deny that democracy exists in Slovakia, and see its tender sprout submerged in a new or revived authoritarianism. Slovakia's image in international media and organizations such as the Council of Europe's Parliamentary Assembly or the European Parliament has deteriorated markedly during the first year of independence, and perhaps more so than that of any other Eastern European country. In what follows, I will shed light on political actors and their strategies as well as on interests, values, and attitudes of social groups, and present the problems encountered in terms of possibilities and choices. Before this can be done, I need to restore them, however briefly, to their broader historical and political context.

Studying political culture under communism

During 41 years of communism building in Czechslovakia, political science as such did not exist, its concerns being taken care of by "scientific communism". During the few pre-communist years , however, a number of academic and other scholarly institutions were established, some of which, including the Czechoslovak Institute for Public Opinion set up in 1946, quickly achieved high professional standards (Wightman and Brown 1979). From those early years, surveys on the popularity of historical political dignitaries important in the history of the Czech and Slovak nations have been preserved. Most importantly, these data reveal what Czechs and Slovaks considered to be the most and least glorious periods in the history of their respective nation, enabling us to compare Czechs' and Slovaks' perception of that history. Such comparisons can be made for the years 1946 (prior to the communist *coup*), 1968 (after twenty years of communist rule and during the Prague Spring), and 1990 (after the collapse of communism). They provide the empirical basis for reconstructing divergent attitudes to the Czechoslovak state on the part of Czech and Slovak citizens.

Eastern Europe knew nothing comparable to Western political sciences' revived interest in political culture during the 1950s and 1960s, partly due to the fact that comparative theory and empirical research crossing national borders did not square with Marxist-Leninist ideology. Only during the short period of "socialism with a human face" was the creation of research institutes experimenting with aspects of Western political culture concepts made possible; at the Institute for Public Opinion, for instance, market analyses and political forecasts were conducted. A number of other research programmes used a theoretical and empirical framework approaching that of political culture, although researchers did not explicitly invoke the concept. Between 1969 and 1989, annual opinion polls and more detailed survey research focusing on political issues took place. But it was concerned only with the Communist Party, with forging evidence of positive responses to government policies at home and abroad, and with finding support for Marxism-Leninism.

Perestroyka breathed new life into survey research during the 1980s, though hardliners still did not permit discussion of such taboo topics as drawbacks of socialism, widespread disillusionment with "communism building" and the growing acceptance of Western lifestyles. Whenever sensitive issues were approached, this had to be done carefully and in a roundabout way – interviewees were asked, for instance, to specify their views on particular aspects of the socialist constitution in force or on particular events such as the Reagan-Gorbachev meeting. One of the more interesting polls was conducted by the Slovak Statistic Office's Institute of Public Opinion Polls (Benkovicova 1990). Its objective was to identify the image of democracy with Slovak citizens. 57% of respondents described "socialist democracy" as any activity not violating the law, 55% thought of democracy as the competition of producers, and for 38% the coinage summoned the activity of private enterpreneurs (which, by then, did not even exist in Slovakia). 22% identified democracy with the rule of the communist party and the working classes, and for a further 11%, it was demonstration rallies and strikes which epitomized the rule of *demos* (respondents were asked to choose three out of ten answers).

Though some of these views seem to have remained unaffected by Slovakia's Velvet Revolution, there are of course marked discrepancies between the findings of 1988 and those of 1989. In particular, only three percent of respondents considered countries belonging to the former Soviet bloc as democratic, as against 20% in the previous year. Only 5% of interviewees believed their own country to be democratic, and even of those the majority were older than 60 years. A significant 15% thought that there were no democracies in the world, and that no such country could ever exist.

Under communism, opinion polls and other surveys results were never published. While the media were fed with fictional data corroborating whatever happened to be the official position, research data were intended for the *nomenklatura* to be used as guidelines for focusing propaganda. The existence of such data also shows that party leaders were reasonably well informed about popular responses to them and their policies. To say that the period of "communism building" was marked by deep perceptual gaps between the communist elite and civil society, therefore, does not seem plausible. The Party knew well what was going on, but was also powerful enough to ignore public opinion.

Slovakia's lack of democratic traditions

Like other post-communist states, Slovakia has a mixed tradition of democracy and authoritarianism with roots not only in the period of communism building, but going back, at least, to the early nineteenth century and the Slovak emancipation movement (Ágh 1994). Ludovit Stur (1815–1856), in promoting Slovak cultural expression and defending his compatriots against Hungarian pressures for assimilation, already addressed the two abiding and critical issues in Slovak national assertion. No doubt the founding of Czechoslovakia as a voluntarily entered federation was the best of all possible solutions for both nations. Though conceived as a state of two small Slav nations, it in fact comprised many other nationalities or ethnic groups. Before the end of World War II, there were more Germans than Slovaks, as well as substantial minorities of Ukrainians and Hungarians. All these groups had their own political parties, and the Communist party was the first to unite them in one political organization. According to the official view of Czechoslovaks as a single nation, Czechs and Slovaks were organized politically as equal partners. But in practice Czechs, or Czech political elites, assumed the paternalist role of stimulating backward Slovakia economically and culturally, and consistently disregarded Slovak demands for more autonomy. Formally, Slovaks' biggest success in wresting control from the political centre was the new federal constitution of 1969. But of course even this limited autonomy never properly materialized, and Slovakia never learnt to take charge of its own affairs.

Neither Czechs nor Slovaks were ever allowed to take their nations' existence for granted. But the many crises affected the two nations differently, and one of the most frequently invoked aspects of Slovak specificity – one used to show the narrow limits of Slovak democratic experience – is the fact that they never had a state of their own.[1]

Communist ideology was aggressively propagated through all means of political socialization. As in all other socialist countries, it fostered a simplistic and Manichean worldview and widespread historical ignorance that seems to have taken hold also of the new intellectual and political elites. A "self service" approach to history distorted and damaged historical as well as political hindsight; Slovakia's collective memory is still ridden with suspicions, figments and scanty knowledge, and is thus an excellent case in point. Here, as elsewhere, communist indoctrination resulted in an historical and normative vacuum (Abraham 1993). Moreover, it is deeply rooted in the minds, value orientations, and behaviour of contemporary Slovaks. We are now reaping the fruit of forty years of indoctrination and are having to face the fact that communist ideology is in some sense still a part of social consciousness, whether of so-called ordinary citizens or of political representatives. The "wall in peoples heads" originally noted in East Germany also besets Slovakia. No matter how genuinely and enthusiastically people welcomed the fall of communism in the streets, for the time being this needs to be seen as an attempt at triumphant negation of the recent past rather than as one of coming to terms with it.

The new Slovak state in the eyes of its citizens

Hardly anyone would have questioned the federations' legitimacy in November 1989. According to opinion polls, there simply was no majority for breaking up the common state. In April 1992, just before the elections leading to divorce, support for the federation was in fact prevalent among adherents of all important Slovak parties, including the soon-to-be governing party led by V. Meciar, the Movement for a Democratic Slovakia (HZDS). Only the Slovak National Party (SNS) demanded an independent Slovak state. A majority of voters wished to see a change in Slovakia's position within the federation, but did not want it dissolved. Accordingly, HZDS's victorious election programme did not call for a breakup but instead advanced five options for a future coexistence with Czechs ("authentic" federation, federation, confederation, union or independent states). This means that neither HZDS nor the Czech Civic Democratic Party (ODS) of Vaclav Klaus had a clear mandate to divide the common state. Both Mečiar and Klaus later rejected the idea of a referendum in the full knowlegde that neither might gather enough ballot support for going ahead with dissolving the federation.

The breakup was generally accepted with resignation on both sides: only a small group of intellectuals initiated protest actions but were unable to mobilize broader support. Some of them set up a spiritual federation wishing to advance cultural exchange despite political separation.[2] In May 1994, only 26.8% of Slovaks claimed that they would have voted in favor of independence had they been given the opportunity to do so, while 57.7% were – in retrospect – against it (FOCUS 1994).

During the first year of independence, critics of the dissolution increased in numbers, comprising former advocates of the common state who had foreseen the negative consequences of the breakup as well as those disappointed with the new Slovak government and its performance in matters of foreign policy, minorities and social policies, and with that government's inability to meet rising social and economic expectations. Increasingly, Slovak politicians became anxious to express their wish to renew ties in some form of union, mainly as a result of pointed gestures of indifference on the part of Czech politicians. The breakup of the federation has been accepted by citizens with a mixture of resignation and disorientation, also within the political elites and with respect to some of their value priorities. There is a slow and somewhat confused process of accepting the Slovak republic as a new reality. Especially in comparison with a Czech republic successful in its main stages of transition, Slovakia displays a marked ambivalence towards independence.

Evaluation of social and political difficulties

The changes taking place in Slovakia since 1989 have been ridden with contradictions. Most Slovaks view post–1989 government and politics generally with a good deal of skepticism readily apparent from the change in labels applied to the collapse of the old regime: the earliest and most poetic and globally accepted epithet "velvet revolution" quickly fell into disuse – only a year later, students began to talk of a "stolen revolution", a "velvet outbreak", a "communist riot", "palace revolution", "Jewish-Bolshevik conspiracy" or simlilarly derogatory apellations.

Dissatisfaction with current political developments seems to be increasingly steadily. In October 1993, 51% of Slovaks saw more disadvantages in the (then) present political system than in the old one, as against 32% in 1991. 12% found no qualitative difference between the two types of systems, as against 22% in 1991. 35% supported the new system, as against 38% in 1991 (FOCUS 1993).

Supporters of the post–1989 regime tend to hold liberal or conservative views in matters of economy and politics, and wish Slovakia to espouse a pro-Western course of development. They constitute a heterogeneous group that may be divided into critics and supporters of Slovak independence, a split running through perhaps all classes, professional organizations, trade unions, and society in general. Positive attitudes toward the post–1989 political system are grounded in two partly contradictory sets of arguments. There is, first, the prevailing liberal orientation emphasizing values of feedom, democracy and individual responsibility in economic and political matters, linked to a desire for greater openness and tolerance. Secondly, support for the present political system comes from a strong conviction that the sovereignty of Slovakia was a necessary and desirable result of the fall of communism. Not surprisingly, an unusually large percentage of respondents holding this view would – in retrospect – have voted for a separation of Czechoslovakia if given the opportunity to do so in a referendum.

On the other side of the spectrum, there is a tendency toward idealization of the old regime, with a corresponding increase in nostalgia for the "good old times" and a readiness to forget the injustices of real socialism. This may be gleaned from the haphazard or ambivalent acceptance of economic transformation, from conceptual confusion with respect to important political problems and ways of solving them, and from growing disillusionment with, or fear of, the future. People seem to miss ingrained "certainties" of life under communism. This somewhat regressive search for solid anchorage is clear from the positive evaluation of the role and activity of the Czechoslovak Communist Party. In May 1994, 31.3% of respondents declared its role to have been a primarily positive one (FOCUS 1994).

Rapid disappearance of traditional certainties, overwhelming and confusing changes in value orientations, and discrepancies between the new and old political cultures have caused something of a culture shock. The severity of this shock has to do both with the threefold nature of Slovak transformation – political and economic changes alongside the creation of a sovereign state – and with stereotypical perceptions deeply rooted in public as well as private spheres; stereotypes which are in part also shaped, or at least reinforced, by simplistic Western responses to "nationalist", "separatist", "Christian", "leftist" and "Eastern" Slovakia (Kusý 1992).

From an individual perspective, such stereotypical perceptions flourish in the "normative vacuum" following the collapse of communism. They are formed against the backdrop of a heritage of communist political mentality, and in a climate of exaggerated expectations concerning the achievement of a "Western lifestyle" whose advantages – though assailed under the old regime – have been well known for many years. But the end of such illusions has come quickly as most Slovaks are facing pervasive economic difficulties.

In the public realm, Slovakia's economic difficulties, the political elite's inability to save the federation, and an increasing number of scandals involving prominent politicians have undermined both popular confidence in leaders as well as the legitimacy of new democratic institutions. Thus, in May 1994, 33% of Slovaks expressed unambiguous confidence in their president, but only 7.9% trusted Parliament, 14% the government, 11.5% trade unions, 20% the Church and a mere 5.4% the Media (FOCUS 1994). A sense of alienation from the "new powers" and low confidence in political institutions can partly be explained by the lack of political experience and by the difficulty of building up a new community identity, but also by hasty decisionmaking on local and national levels and by the manifest absence of any conceptual framework for timing system change or for establishing priorities among its many facets.

Remarkable cynicism has followed "velvet" enthusiasm. In 1993, 89% of Slovaks thought that politicians first and foremost took care of themselves and their clientele. 79% were convinced that familial ties, nepotism, opportunism and careerism were what mattered in politics[3] Similar views are expressed in the adage "the rich buy up democracy, they always have done and always will". Thus, attitudes complementary to old regime politics have once again come to dominate Slovakia's political culture: "politics", this wisdom has it, "is a dirty business".

Attitudes toward the market economy

Slovaks' perception of the relationship between state and individuals is dominated by collectivist and paternalist attitudes, and the recent severe economic difficulties have done little to ween Slovaks from this tradition. As a result, demands for protective intervention by the state is still strong: according to a survey carried out in 1993, 89% of respondents thought that the state should give a job to anyone willing to work, 80% wished to see the state more actively involved in price fixing, and 75% expected the state to provide housing for each family encountering difficulties in their search for a place to live (FOCUS 1993). One year later, 49% of Slovaks expressed preference for a "social market economy" in which the state plays a leading role, 14.7% longed for a socialist economy, and 26.9% for a free market economy with minimum interference by the state (FOCUS 1994).

But Slovak opinions on economic matters are not devoid of inconsistencies: far-reaching interventionist expectations often coincide with an endorsement of free market ideals: 64% believe that private enterprise should be given as much freedom as possible, and that even very high incomes of successful entrepreneurs are justified. Foreign companies operating in Slovakia should, according to 65% of respondents, have complete freedom of action (FOCUS 1993). In another survey, however, 54% of respondents maintained that Slovak property should not be sold to foreign investors, a flagrant contradiction perhaps indicative of the high degree of ambivalence and economic illiteracy prevailing in a society nurtured on an all but unbroken tradition of paternalism.[4]

Political elites in transition

Following parliamentary elections in 1990 and 1992, a heterogeneous composition of political forces with no stable party structures, party discipline or clearly articulated political agendas came to power. The aims of many political parties changed with every new issue, and the proclaimed political profile of a group was not usually substantiated on the level of practice. A pervasive "splitting syndrome" affecting political movements of anything approaching broader support, as well as strongly adversarial and competitive behaviour among elites, have created a deeply fragmented political scene. Except for the communist successor party SDL, all major parties have been devided at least once since 1989. There is no clear set of criteria for locating the cleavages separating political elites, but some general rules may be inferred from half a decade of democratic experience: in the main, it has been either personal animosities between movement or party leaders, disagreement over the question of Slovak independence, or the competition for power irrespective of political creeds or party lines that have proved divisive. Unsurprisingly, some Slovak politicians have changed their political affiliation and affinities many times since 1989.

A remarkable feature of Slovakia's political system has been the extreme fragmentation of right wing forces, which was largely responsible for the electoral desasters of 1992. Among the self-styled tribunes of November 1989, excessive ambition and occasionally arrogant behaviour have done nothing to inspire confidence in these parties' problem solving competence.

The replacement of old ruling elites, begun with the intention of barring senior communist officials from office, has also revealed a contradictory pattern. Initially, representatives of "anti-political politics" clung to the requirement for new political leaders not to have been members of the former Communist Party or other organizations belonging to the "National Front" of Communist Party supporters. At the same time, the demand for reintegration of experienced and not directly incriminated professionals soon arose, and recently some prominent former CP protagonists have made comebacks within the political elite. In contrast to the Czech Republic, in Slovakia the often repeated assertion that "there is a lack of competent people" has now led to widespread acceptance of a "recycling model" of elite recruitment. All parties have been affected by these contradictory pulls in their recruitment, which is also evident from the high number of National Council delegates with a political record under the old regime. Most of them, incidentally, have adopted November 1989 as their "political date of birth", disavowing their dealings with the previous regime as well as, it seems, their age.

In business and industry, there has been even more continuity of this kind. Key positions are usually still held by the same managers due to their all-important personal connections at least as much as in recognition of their competence. Members of senior industrial management often raise sufficient capital to start private or privatized companies. Where attempts at rehabilitation have been unsuccessful, such individuals act behind the scenes and through relatives or confidants.

Since Slovakia's media are open to influence peddling and other forms of manipulation, the impact of this smooth transition within the political elites should not be underestimated. Especially in electronic media, persistent and more or less open attempts to wield influence have been made by just about anyone with a political mandate.

Social and political pluralism

In contrast to other socialist countries of the 1980s, the Czechoslovak regime did not tolerate any independent organizations or vaguely political activities. While Czech dissident groups managed to circumvent repression to some degree, attempts at spreading their activities and creating a "parallel society" (Kusy 1990) in Slovakia failed, apparently as a result of effective law enforcement and widespread fear of reprisals.

The Velvet Revolution established constitutional rights as well as the legal framework for civic and political associations. Under the pressures of liberalization, marketization, and the ensuing rapid social differentiation, Western organizational and institutional patterns quickly came to dominate the otherwise slow rebirth of civil society. Again, the reaction towards them was ambivalent. They were, on the one hand, accepted as yet another prestigious Western import, but were too reminiscent of National Front – the communist umbrella for all interest and other groups supportive of the CP – with respect to structures, professionalized leadership and even activities. Membership in professional organizations is, however, widely accepted, quite in contrast to the wholly discredited notion of joining or participating in a political party. Many *de facto* parties have recognized this aversion by posing as movements, forums, alliances, or unions. It is important, however, not to confuse the decreasing "participation in politics" with "interest in politics". While for students interviewed in 1993 participating in politics through established institutions lost out against fifteen other "favourite activities",[5] their interest in politics as well as their readiness to participate in more informal setups and activities does not justify their categorical description as "apolitical". An element of continuity with institutional practices under communism may be seen in the tendency of a minority of young people to join parties as a first step in career management.

The media

The advent of pluralism in print and electronic media has been a decisive factor in system change, even though the television monopoly has still not been abolished despite a heated public debate and numerous applications for private broadcasting licences. There have been many attempts on the part of the political elite to curtail the independence of the media, all based on the conviction, in part borne out, that control of the mass media equals victory at the polls. Following a similar logic, some newspapers are owned by, or affiliated with, political parties, such as the daily Pravda of HZDF, or Slovensky Dennik of KDH.

Religion and churches

Slovaks show relatively high levels of religiosity (82.5% of its inhabitants declaring themselves as believers) with a majority of Catholics (roughly 60%) and 6.2%

Protestants. 9.7% consider themselves as atheist (Cenzus 1991) But political influence of, particularly, the Catholic Church was rejected by 63% in 1993 (FOCUS 1993) Thus the fear of political Catholicism is shared by protestants, atheists and, though to a lesser extent, catholics alike. Reasons for this may be found in the alliance formed between catholicism and fascism during World War II, and in the Church's attempt to push for tighter abortion laws. Many Slovaks also resent the Church's intention to extend censorship to, for example, sexually explicit television programmes.

The national dimension in political culture

One of the key issues in Slovakia's transformation is undoubtedly the question of whether and how to reconcile civic-democratic and national aspirations. Within Slovak nationalism, two different traditions may be distinguished – one more conservative, inward-looking, and populist, the other more open, assertive and sanguine – which still largely set the terms of debates over Slovak nationhood. Unsurprisingly, Slovakia's numerous minorities react with particular sensitivity to the discourse of national assertion.

From its inception, Czechoslovakia comprised many minorities, most of whom concentrated on Slovak territory. Today, roughly 15% of the population declare themselves as members of one of the following recognized national minorities: Hungarian, Ukrainian, Rusine, Romany, German, Polish or Croation. These are variously dispersed or concentrated in enclaves across the whole of the country. Jews, while in many respects a minority, are not officially recognized as ethnically distinct and have so far escaped statistical measuring. Following the "velvet divorce", a new minority appeared in either state: Slovaks in Czechia and Czechs in Slovakia are only beginning to get to terms with their new minority status, opting for one or the other citizenship and searching for ways of lending expression to their collective identity in changed surroundings. Various associations and initiatives as well as publications have accompanied that search. In August 1994, a General Assembly of local Czech organizations took it upon itself to "promote the development of social, cultural and academic life of Czechs, Moravians and Silesians in Slovakia" (SME 1994).

Especially the presence of Hungarians reveals the complexities of accommodating minorities in its national and cross-national dimensions. A language law adopted in 1990 provides for the use of Hungarian in all official matters on a local level, if more than 20% of a community declare themselves Hungarian. On Slovakia's acceptance into the Council of Europe, spelling and grammar rules concerning womens' surnames were adapted and provisions for bilingual road signs made. Relations with neighbouring Hungary, strained as they have been since the quarrel over a hydroelectric dam at Gabcikovo, are suffering further under a permanent dispute over the national composition of Slovakia's regional districts.

Minority rights have, of course, become heavily politicized. There is a sense that relations between Slovaks and Hungarians are concerns mainly for those unaffected by them in their daily lives, that conflicts are overly dramatized and abused for other political purposes (Fric 1994). As everywhere, resentment against Hungarians flares up mostly in regions where none are to be found. In Parliament, two parties and a cross-party "Hungarian coalition" represent Hungarian interests.

For the first time in sixty years, Romanies could openly declare their ethnic affilitation in the census of 1991. Only 1.52% of Slovak citizens did so, however, although Romany representatives and government officials estimate that between 8% and 12% could have done. Though granted minority status and legal recognition, as well as some support in developing ethnic and cultural self-awareness, the social situation of Romanies is still precarious: they occupy the least educated, poorest, and most frequently unemployed sectors of Slovakia's population. Resentment against Romanies can be found across all distinctions of class, profession, education, gender, age, political preferences and religion in Slovak society. A relatively high crime rate among Romanies commonly serves as justification for the idea that the "Romany problem" should be dealt with by means of repression and isolation (FOCUS 1993). Quarrels among Romany representatives – who are divided on such fundamental issues as whether Romanies should claim separate ethnic status or rather pursue a strategy of rapid social integration – have not helped to either relieve their social problems or erode the stereotypical and negative projections confronting them (Mann 1992).

Ethnic relations in Slovakia are thus a source of potentially disruptive conflicts within the country as well as with its neighbours, especially in times of colliding and forcefully advanced ethnic and national claims. Nevertheless, in 1994 only 5% of Slovaks identified ethnic struggle as a possible source of serious conflict (FOCUS 1994).

Conclusion

A majority of citizens in Slovakia, as in other post-communist states, firmly rejected communism in its principles and practices by the time it fell apart. Although some may be opposed to the new direction taken by their country, this does not amount to saying that they are or will ever be proponents of a return to the old order. Whatever disagreements or variant interpretations there may be, Democracy and market economy are ultimately the shared objectives of almost all political actors. Against this backdrop, crises and setbacks are not quite what democracy's "prophets of doom" make them out to be. I remain convinced that Slovakia's development will confirm these conclusions, and that its citizens will never again be mournful survivors wailing over the grave of democracy.

Notes

1   Except, that is, from 1939 to 1945 and under a Nazi puppet government. In 1945, Slovakia joined the allies only thanks to a revolt against Jozef Tiso's People's Party staged and crushed in 1944.
2   'Report on the Czech and Slovak Republics', International Institute for the Study of Politics, vol. 6, June 1994, Brno.
3   The Role of Political Culture in the Transformation of Post-Communist Societies', French Centre for Research into Post-Communist Societies, CNRS and Department of Political Sciences, Comenius University, Bratislava, 1993
4   see note 3.
5   see note 3.

References

Abraham, S. (1993), *The Break-Up of Czechoslovakia: A Threat to Democratization in Slovakia?*, Ottawa.

Ágh, A. (1994), 'The Revival of Mixed Traditions: Democracy and Authoritarian Renewal in Eastern Europe', paper presented at the Conference on Democratic Modernization in the Countries of East and Central Europe, Essex University, May 1994.

Benkovicova, L. (1990), 'Obraz demokracie vo vedómi abčanov Slovanska' (The image of democracy in the minds of Slovak citizens), unpublished research report, Institute of Public Opinion Polls, Statistics Office of the Slovak Republic.

Cenzus (1991), Scitanie ludi, domov a bytov v ČSFR, Prague.

FOCUS, Center for Social and Market Analysis (1993), 'Current Problems of Slovakia after the Break-Up of the ČSFR', Bratislava.

FOCUS, Center for Social and Market Analysis (1994), Opinion Polls conducted from 9 May to 23 May 1994 on a sample of 2018 Slovak respondents

Frič, P. (1994), 'Myths and realities in Southern Slovakia', *SME,* 5 August.

Kusý, M. (1990), 'The Slovak Phenomenon', *Fragment K*, no. 3.

Kusý, M. (1992), 'Slovaci su viac...' (Slovaks are more...), *Slovenské Pohadly*, September.

SME (1994), (Slovak daily newspaper), 8 August.

Mann, A. B. (1992), 'The Formation of the Ethnic Identity of the Romany in Slovakia', in *Minorities in Politics*. Czechoslovak Committee of the European Cultural Foundation, Bratislava.

Wightman, G. and Brown, A. (1979), 'Czechoslovakia: Revival and Retreat', in Wightman, G. and Brown, A. (eds.), *Political Culture and Political Change in Communist States*, Macmillan, London.

12 Political Culture in Slovakia. Populism and Nationalism in the Context of Popular Values

Gregorij Meseznikov

Post-communist transition is a complicated process of development in economic, political, and socio-cultural subsystems of society. Its general direction, portent and depth depends on the kind and degree of interaction between these subsystems, on the synchronicity or asnchronicity of processes taking place within their frameworks, and on their relative autonomy. According to the Slovak sociologist Soňa Szomolányi (1993), a defining feature of Slovak post-communist transition has been the fact that

> the economic subsystem is still not sufficiently autonomous. Reforms within the political system must therefore provide a favourable basis for economic restructuring, which is why they are overwhelmingly relevant as factors of change. But political subculture is in turn largely dependent on developments within the socio-cultural subsystem. Among the three subsystems mentioned, the political is the most changeable in Slovak society, while the socio-cultural system and the background of normative values it creates for economic and political life shows the highest degree of stability.

Political culture thus plays an important role in the process of interaction between political and socio-cultural spheres. Its mediating function stems from the fact that its parameters of value are set within the social context of these values. It denotes a particular attitude to political matters which is mirrored, to some extent, in the structures of the political system. Its significance as a concept lies also in the fact that it allows for research on the deeper causes of a given group's or individual's political behaviour. The concept of political consciousness is derived from that of political culture, and denotes an individual's perception of political reality and the actions prompted by that perception.

We can distinguish between mass, group, and individual political consciousness. Mass political consciousness itself includes both relatively stable value orientations and dynamic components, such as

1. a population's social expectations,
2. the evaluation of prospects of realizing these expectations,
3. socio-political values shaping ideological preferences (democracy, equality, liberty, stability, order etc.),
4. swaying moods and popular opinion concerning parties, movements, leaders etc.

Mass consciousness thus conceived determines a society's political culture as well as the viable variants of political behaviour. On the level of groups, we consider collective identity insofar as it is shaped within concrete social entities, such as associations, strata, ethnic and religious communities, political organizations, and others. Finally, on the level of individuals, political consciousness describes the *zoon politikon* perceiving, evaluating, and acting upon, the political process.

Research on these stable and dynamic components of political consciousness should allow for relatively reliable prognoses concerning political development during transitions from totalitarianism to democracy. Provided the three levels of analysis are carefully distinguished, it should also be possible to narrow down future possibilities and contingencies of political development to a relatively few, relatively probable ones. In Slovakia, some scholars have embarked on such research on the basis of data gathered by FOCUS (Center for Social and Marketing Analysis). In what follows, I rely on the work of Soňa Szonolányi, Vladimír Krivý and Tatiana Rosová.

Basic value orientations of the population are of primary importance for any diagnosis of Slovak political culture. Perhaps the key question to be asked in these times of transition is the one concerning public attitudes toward the old regime. According to survey research carried out in 1992, only 50% of Slovaks were convinced of the need for radical social and political transformation. The other half expressed preference, though in retrospect, for communist government, probably in response to economic hardship accompanying structural change. 50% disapproved of any belt-tightening measures, suggesting that dissatisfaction with the new political institutions is in the main economically motivated.

While Czech citizens predominantly feared the failure of transformation, Slovaks seem to have been more afraid of its consequences. This divergence in attitudes can be understood when seen against the backdrop of equally divergent patterns of modernization in the two countries. In Slovakia, the so-called "overtaking" modernization of the socialist type was instigated after World War II. It entailed heavy influence of the state on economic structures, social composition of the population and socio-cultural parameters. A large part of Slovakia's population saw political development following November 1989 as a loss of previously guaranteed social standards. According to V. Krivý, distrust in transformation has even become dominant, an interpretation substantiated by the results of the 1992 parliamentary elections: the Movement for a Democratic Slovakia's victory was mainly the result of its outspoken criticism of "federal" or "radical" economic reform.

In Slovakia, the question of social and economic transformation was inextricably linked to the issue of Czecho-slovak federation. The Movement for a Democratic Slovakia (MDS) managed to fuse distrust in reform with wariness of Czech dominance and uncertainty regarding the future of the federal state. This resulted in a powerful momentum, by means of which the elite of MDS not only rode to election victory but also contributed significantly to the ČSFR's breakup.

According to Vladimír Krivý, contemporary Slovak society fits the following descriptions:

1. High expectations regarding the performance of the state. More than 50 per cent of Slovak citizens support far-reaching intervention in the economy. In 1992, 41% favoured a so-called mixed economy combining market and socialist elements.

2. Marked egalitarianism. Two thirds of respondents favoured steps toward reversing income disparities, a view often coupled with pervasive distrust of individuals successful in the private sector.

3. Considerable economic utopianism. The inability to weigh the importance of macro-economic trends, advocacy of contradictory economic strategies, or susceptibility to absurd populist promises are typical of such views.

4. Prevalence of passive or despondent attitudes regarding the shaping of one's life.

5. Relative weakness of liberal values.

6. Susceptibilities to strong leaders. In 1992, 58% of respondents preferred strong leadership to the politics of negotiation and compromise. In 1993, however, this trend toward authoritarian solutions was reversed somewhat.

7. An inclination (37%) to a view of democracy as majority power without respect for minority rights.

According to Tatjana Rosová's analysis (1992), there are seven major discernible groups within Slovak society and its political allegiances.

1. Liberals (15%) are in the main convinced of the necessity and directedness of economic reform, reject state paternalism and stress values of personal responsibility.

2. National-liberals (11%) accept the principles of parliamentary democracy, reject authoritarian leadership, and (prior to the velvet divorce) saw Czecho-slovak federalism as unfavourable to the latter without harbouring strong anti-Czech feelings.

3. Federal democrats (6%) display positive attitudes towards Czechs, the Czech republic and (again, prior to its breakup) toward the federation. They also show the highest levels of respect for parliamentary democracy and its institutions.

4. Nostalgics (11%) typically distrust democratic governance, express preference for a centralized, state-owned and state-run economy, and approved of the federation.

5. Non-liberal nationalists (15%) show a high degree of intolerance, xenophobia, anti-Hungarian, anti-Czech and antisemitist attitudes, as well as preference for state ownership.

6. Radical anti-liberals (15%) are similarly intolerant towards ethnic and other minorities, and reject democracy and economic reform alike.

7. The wavering and indifferent, who amount to about 15% and express no clear views on major social and political questions.

In order to know how such differences in value orientations affect the development of politics, we need to look at how they are expressed on the level of political formation. Slovakia's political scene is remarkably dynamic. Permanent regrouping and realigning among political forces is the rule. This is part of a general search for political identity that may be seen both as a consequence of, and response to, discontinuous development in earlier periods of the country's history. In order to conceptualize this fluidity, it is perhaps more helpful to characterize Slovakia's political scene not as one dominated by a "left" and "right" divide, but as a political spectrum between the extremes of paternalism and liberalism.

The pattern of paternalism and the desire of many people to melt into impersonal entities like the state, nation, ethnic group, or religious community is still the prevailing one. This is why even a change in political allegiance (such as support for a different political party) does not necessarily imply an alientation from paternalist values. Voting for a different party is compatible with such continuity, since there is a paternalist segment or faction emphasizing social, national and religious aspects in this order in almost every larger political association. The primary goal of social transformation must therefore be to question effectively the paternalist paradigm and encourage more and more people to abandon it. Only then would political participation occur with a desirable minimum of personal responsibility and priority for civic virtues, or an the basis of a – broadly conceived – liberal paradigm.

The marked personification of political life is typical for Slovakia. Attitudes of pragmatism have, as yet, hardly entered the political sphere, and issues are consequently decided all but exclusively by party allegiance. Distrust in the emerging institutions and a general sense of unease or uncertainty accompanying the transformation process put even stronger pressure on prominent individuals to win public confidence and centre a party's profile around themselves. A high degree of voter fluctuation shows that Slovakia's political scene has so far not sufficiently crystallized or stabilized. Since 1992, at least fifty percent of voters have changed their party preference. Mostly, votes flowed within the triangle defined by the Movement for a Democratic Slovakia (MDS), the Party of the Democratic Left (PDL), and the Slovak National Party (SNP). The search for a suitable representative was particularly intense on the Left, and it is tempting to relate this phenomenon to the broken traditions of parties and orientations caused by aggressive post-war homogenization of social structure. Only the communist successor party PDL and, to some extent, the Christian Democratic Movement (CDM) enjoy a more stable constituency rooted in shared beliefs.

Mutual exchangeability of beliefs deeply affects the forming of party preferences. During periods of regime change, especially supporters of populist and authoritarian mass movements approach political issues with a set of disparate but convertible ideologies. Reasons for this should be sought not merely in party rhetoric and election campaigning, but also on the level of value orientations. Thus voter flow between CDM and MDS from 1990 and 1992 seems to have been motivated primarily by persistent collectivist and paternalist preferences.

With respect to value orientations of different parties and their supporters, Slovakias political scene is differentiated internally, that is, organized around a political and an economic axis. Authoritarianism and liberalism represent either end of the political spectrum, with principles of order, national unity and power accompanying the one, and ideals of personal freedom, ideological plurality and political pragmatism the other. Supporters of the Movement for a Democratic Slovakia (MDS), the Slovak National Party (SNP) and similar groups are predominantly authoritarian, while those of the Christian Democratic Movement (CDM), the Hungarian Christian Democratic Movement (HCDM), the Democratic Party (DM), the Entrepreneur's Party (EP), the

Hungarian Civic Party (HCP), the Democratic Union (DU), and the National Democratic Party (NDP) are closer to the liberal end. Economically, however, the advocacy of marketization on the one hand, and of state dominance on the other, oppose PDL, MDS and DP, EP, HCP, CDM, DU, NDP respectively. But the *de facto* differences between, say, supporting PDL's left wing and supporting a right wing party in matters of economic policy are not as significant as might be expected on the basis of these parties' political profile. In part, this is due to the fact that the implementation of economic reforms has not been accompanied by political differentiation, and that, as a consequence, there is no very clear articulation of the divergent interests in East Central European societies generally. Widespread inconsistency in the formation of policy preferences are the result. Even supporters of right-wing parties usually favour state regulation of prices and labour markets, indicating again that the formula of "paternalism vs. liberalism" may be more useful than one distinguishing between "right" and "left" politics. It is also easy to see that the "leftist" label addresses paternalist attitudes across the spectrum, including those of people with right-wing party preferences. Although it is possible to say that state-paternalist views decrease toward the right end of the spectrum, they are still relatively frequent even among supporters of CDM. Thus, even for these voters, economic policy is not a decisive criterion in the choice of a particular party. Patterns and correspondences must be sought elsewhere, and they are most likely to be found by linking political behaviour with Slovak socio-cultural parameters.

Populism

The role and influence of populist politics in Slovakia and other post-communist societies has recently been a hotly debated topic. Populist movements and their leaders of course proclaim unanimously that their main goal is the welfare of the people, and their popularity accordingly varies with the severity of economic, social, cultural or political crises. Advocacy of "traditional" and "ordinary" popular values are typical of such movements, and the "simple folks" are invariably the main subject of populist theorizing. The apparent representation of their interests by means of "popular queries" have become a main strategy of Slovak populists, since they are a means of advancing simple solutions to complex problems. Populists do not merely express themselves in a language simple enough to be understood by large audiences; they also choose their topics in accordance with what they take to be popular priorities. In their approach to current issues, populists bring up aspects of (national, ethnic, religious, racial, ecological) social reality much more agressively than most of their rivals in the political field. A typical combination of the above strategies was the assertion, made during the 1992 elections, that it would be possible to solve all problems of Slovakia's socio-economic development by simply liquidating the federal Czechoslovak state. Similarly, populists often capitalize on group distinctions in order to win the favour of the currently larger, or more clamorous, or otherwise politically more expedient. To win over the rural population, for instance, they praised the moral superiority of ordinary villagers over depraved urbanites. In order to obtain the support of "the great masses of working people", populists vaunted their being exposed to and familiar with the real problems and needs of society, in contrast to "sophisticated and pampered intellectuals".

In short, anyone with a convincing claim to popular roots or to understanding the ordinary citizen is well placed to prevail in the politics of Slovak populist mudslinging.

Liberal and national paradigms

The relation between liberal, democratic values on the one hand, and "national" ones on the other, is another crucial aspect of transition in post-communist societies, and especially so for multi-ethnic countries like Slovakia. After decades of modernization in conditions of unfreedom, the values of liberal democracy have become so attractive to ECE societies that the period immediately following the collapse of communism may be desribed as one during which a basic liberal consensus was formed. This was a consensus initially underwritten by most social forces, even though subsequent development showed that some of them had rather little in common with even a broadly conceived liberalism. Political representatives of diverse orientation therefore all began to acquire public profile by starting from the common ground of essentially liberal principles. At the same time, however, ethnic and national preoccupations began to be expressed and to constitute a disintegrating force. Some European post-communist states had been internally divided into national communities (Soviet Union, Yugoslavia and Czechoslovakia) which the collapse of communism turned into sovereign states. As we all know, the appearance of these new states led to ethnic conflicts ranging from disputes to the bloodiest of wars. Whatever the intensity of such conflicts, they have complicated bilateral relations wherever they occurred, and have also presented serious challenges to the process of anchoring liberalism and democracy. We know already that in some cases, the challenges of nationalism have proved fatal for the continuity and eventual success of social transformation. In search for adequate responses to such challenges, those striving for an open society will need to ask themselves what choice of action liberalism and democracy have when faced with roused feelings of nationalism, what kind of strategies are viable and necessary if a nascent civic society is to withstand the pressures of growing ethnic and national particularisms.

The task of securing democracy is complicated by the fact that in some countries, those trying to contribute to civic society find themselves outnumbered by those undermining it. This is largely a result of long term trends in social structure, of political, religious and cultural traditions, of communist social engineering and the specific value orientations all these factors have promoted. We need to, first of all, find out what the role and function of nationalism in contemporary post-communist societies is. Nationalists typically affirm what they consider to be the individuality of a nation, asserting that individuality can take many forms but ultimately aims, of course at national self-determination in a sovereign state and on a secured territory. Slovak nationalists thus argued that Slovak individuality was systematically oppressed. According to Isaiah Berlin (1969), an individual could feel oppressed not just because he or she is not recognized as an individual, but also because he or she is a member of an unrecognized social, racial, national or other group. In some sense, the values of national liberation could therefore be compatible with, or seem like a natural extension of, basic "liberal values" of personal freedom (though in reality there is hardly any discernible relation between the freedom of the nation and that of individuals). With respect to Slovakia, it is questionable at the very least whether the majority of citizens actually thought of themselves as part of a non-recognized national or ethnic entity.

Neither opinion polls nor qualitative analyses have confirmed the hypothesis that most Slovaks saw the federation as oppressive. The advent of independent Slovakia, which is nowadays interpreted as the culmination of a long struggle for national emancipation, was thus not the consequence of a struggle with particularly wide or intense participation. It was not accompanied by anything approaching the emerging and powerful national movements we know from the history of decolonization, suggesting that a select few powerful interests formed the backbone of the supposedly "national" movement.

Quite contrary to the assumption of a craving for national liberation, an analysis of manifestos, other documents and the political practice of nationalists in Slovakia and other ECE countries indicates that the national paradigm has been a reaction against modernization and transnational economic as well as political cooperation, a defense of obsolete structures and traditions as well as of non-productive social and political patterns (Havelka 1991). "Nationalism is, first of all, a brake in development. It stimulates anti-modernization trends" (Michnik 1994). Despite the claim to address real and pressing social problems, nationalists do not have the capacity or disposition to do so. They glorify, first of all, a national past or future, and have little patience for the vexing realities of the present.

The nationalist upsurge in post-communist societies is connected with the fact that communism destroyed civil society where it existed, or prevented its emergence where it had never even been created. Communist government depended for its relative viability on severing "natural", that is, social, professional, cultural, spiritual, religious and other ties important for the shaping of social identities. Only ethnic ties remained untouched by this onslaught, probably because they shared with communist ideology the all-important collectivist element. The common features of egalitarianism and paternalism were shared by the two ideologies and readily found in the countries where they took hold.

After the collapse of communism, Slovak public debate was passionately preoccupied with the possibility of harmoniously combining civic-liberal and national traditions. The first demands the creation of a modern constitution and political institutions within the federal framework. Nationalism, on the other hand, strives for coinciding ethnic and political boundaries, which implicitly calls for an ethnically cleansed state. While Slovakia thus urgently needed to reposition itself within the federation, it was at the same time the scene of increasingly acerbic nationalist discourses, its territory being the home of many larger (Hungarian) and smaller minorities. The political field was soon polarized between, on the one hand, those groups who had gained merit and democratic legitimacy through their contributions to ending totalitarian government and almost unanimously opted for the civic principle, and, on the other hand, those who made "national interests" their banner and were associated with the regressive, nostalgic side of Slovak politics.

Perhaps the issue may be best approached by invoking the old distinction between individual and collective rights. Nationalists's arguments were typically ridden with double standards, insisting as they did on the protection of individual rights in order to counter the claims of minorities, and simultanously championing the collective right of self-determination for Slovaks within the federation. This was so despite the fact that the federal constitution provided for such rights, for example by ruling out majorization in the federal assembly. But the issue of collective rights has been further complicated by its confusion with the question of the Slovak state's territorial integrity, as well as by a chronic lack of communication between Slovakia's political elites and minority representatives.

The problem of minorities has, of course, much wider implications for the quality of democracy, and this is especially true of fledgling democracies such as those of East Central Europe. Minority rights have become highly sensitive and crucial issues, since they raise the question as to whether, and to what extent, human rights can be made to work. They have become the "touchstone for the concept of modern democracy" (Kusý 1994). In traditional Western democracies, collective rights or prerogatives of minorities have in the main been balanced with individual rights in such a way as to compensate for disadvantages which members of minority groups almost invariably have. These solutions are "liberal" insofar as they try to redress the inequalities of real life. To root this understanding in political culture seems important because "cultural feedom and pluralism at present are almost certainly better safeguarded in large states which know themselves to be plurinational and pluricultural than in small ones pursuing the ideals of ethnic-linguistic and cultural homogeneity" (Hobsbawm 1992).

For other reasons, too, nationalism is inimical especially to nascent democracies. First, its concept of democracy is that of majority rule, and correspondingly shows contempt for representative institutions, division of power, and constitutional pluralism. It will try to enforce homogeneity and reject the idea that securing the interests of all includes those of minorities. It will also tend to advocate a concentration of power and strong authoritarian leadership.

Second, nationalism opposes liberal democracy with its propensity in favour of *étatisme*, proclaiming the priority of the state over any other form of social organization within ethnic commmunities. It sees the creation of the state as both the first and most important political objective; along with other paternalist, egaliarian and collectivist mindsets, such *étatisme* is thus a major obstacle to the fostering of open societies and responsible citizens. There is a permanent danger that some political representative may succeed in convincing a large number of followers that he or she, and, by implication, they, represent, embody, or simply are the state. In conjunction with widespread dependency on the state and its diverse provisions, such leaders suggest that citizens have the choice to rally behind the "wishes of the nation" or be left out and face ostracism.

Third, nationalism has shown little inclination towards peaceful coexistence, neither internally nor externally. Its view of representative democracy does not easily lead to compromise and tolerance. For nationalists, their creed is the "obvious principle" shrugging off any relativization by definition.

What chances, therefore, do liberal and democratic forces have in a country which has come under nationalism's sway at the outset of transformation, and will probably see more ethnic particularisms and strife in the nearer future? What hope, in short, is there for civic and civil politics to prevail over their foes? The main precondition for defusing ethnic particularisms will be a successful transformation decreasing levels of frustration as well as opportunities to exploit them. The more reforms yield concrete and perceptible results, the fewer chances for nationalism.

Primary importance should be attached to creating and defending a *Rechtsstaat* with which to stabilize the workings of democratic institutions. Liberally and democratically minded intellectuals would perhaps do best to try and close communication gaps (Gellner 1983) between ethnic groups and their elites, and to reveal the consequences of populism and its political practices

To end on an optimistic note, there is now evidence that a majority of Slovaks are critical of the wave of "post-November" nationalism (Šimečka 1994). On this basis, it should and must be possible to erode stereotypes and, in time, to create an atmosphere of tolerance in which diversity can flourish.

References

Szomolányi, S. (1993), *Tranzicia slovenskej spoločnosti: explanačný rámec*, Filozofická fakulta UK, Bratislava.

Krivý, V. (1993), 'Socio – kulturne pozadie transformacnych procesov na Slovensku' in *Slovensko: kroky k Europskemu spolocenstvu. Scenar socialno – politickych suvislosti do roku 2005*, Sociologicky ustav SAV, Bratislava.

Rosová, T. (1992), *Aktualne problemy eskoslovenska – Januar 1992*, Centrum pre socialnu analyzu, Bratislava.

Hobsbawm, E. J. (1992), *Nations and Nationalism Since 1780 : Programme, Myth, Reality*, Cambridge.

Berlin, I. (1969) 'Two concepts of Liberty', in *Essays on Liberty*, Oxford.

Gellner, E. (1983) *Nations and Nationalism*, Oxford.

Havelka, M (1991), 'Narod jako posvátný statek', *Přitomnost*, no. 11.

Michnik, A. (1994), 'Demokracia môže aj umriét Sme na nedélu', 3.2.1994

Kohn, H. (1955), *Nationalism. Its Meaning and History*, Princeton.

Kusý, M. (1994), 'Kolektívne práva?' *Listy*, no. 2.

Šimečka, 'Narodni neprirozenost' *Listy*, no. 2.

CZECH REPUBLIC

13 Czechs and Slovaks after the Velvet Divorce

Pavol Frič

Czech and Slovak views on the general economic situation

Three years have gone by since the Czecho-Slovak divorce on 1 January 1993. Prior to that date, anyone giving the Czechoslovakia's breakup a serious thought must have worried about the possible impact of this move on economic and political development in the new countries. Two extreme scenarios widely credited even by experts were painted – one, more idyllic, for Czechs, the other, catastrophic, for Slovaks. A Czech ship shedding the Slovak burden, it was said, would rapidly catch up with Western countries, while a Slovak train without a Czech locomotive would soon grind to an economic halt. An estimated exchange rate of one to three for Czech and Slovak crowns respectively seemed to shed a dramatic light on such prognoses. But only a year later, some pointed out that the divorce would hit both countries economically, no matter how smooth the separation. The state of the Slovak economy did not seem quite so catastrophic, and there were even those mesmerized by the vision of a Danube eldorado or "Slovak Switzerland."

This debate among economists has continued to the present day. Some of them may be found vaunting "fantastic macro-indices" of the Czech economy and raising a warning finger at faltering Slovak reforms, while others see those reforms as progressing normally and point out that living standards of ordinary people in the Czech Republic are actually deteriorating. But whatever their methodological and theoretical repertoire, economists are not usually found among those actually struggling for survival. For a closer understanding of political culture and its likely development, it will therefore do no harm to look at the situation from the perspective of an – however imaginary – ordinary Czech or Slovak citizen.[1]

The views of CR (Czech Republic) and SR (Slovak Republic) citizens on the economic situation of their country have changed significantly over the last two or three years. It is possible to distinguish the following three stages in the development of these opinions.

1. A period of general agitation in 1992, when parliamentary elections took place and most people were just coming to terms with the reality of Czechoslovakia's imminent break-up. This time was characterized by widely fluctuating opinions on the economic situation.

2. A time of optimism and faith in the future of the newly independent Czech state, which lasted approximately until the beginning of 1994. The fear of possible negative consequences of the ČSFR's dissolution quickly subsided, and the economic viability of the new state ceased to be doubted.

3. A period of disillusionment or simple fatigue with waiting for more palpable effects of economic prosperity. This stage is not so much characterized by pessimism as by the widening perception of a lack in tangible and positive changes, resulting in the apparently contradictory coincidence of a slightly improved average rating[2] of the economic situation with "the average citizen's" convictions that things have remained more or less unchanged. This average opinion, however, should not be equated with a majority belief, since it is merely the statistical result of the fact that, in 1995, roughly one third of Czechs felt that the economic situation had deteriorated, while another third (30%) thought it had improved and yet another third (31%) saw no change at all (6% declining to comment).

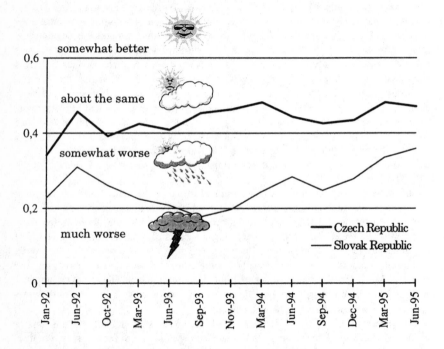

Figure 1 Ratings of the current economic situation

In the Slovak Republic, changes in opinion on the economy have been more difficult to plot. Here, a short period of optimism prior to the elections of 1992 was followed by a persistent rise in negative ratings dropping to the category of "much worse than in the year before" (i.e. prior to the dissolution of the ČSFR). However, this unfavorable trend was reversed beginning with Vladimir Mečiar's term in office as prime minister in Autumn 1993. It was a radical change, though one not so much caused by a proliferation of optimists as by a decline in pessimists and a proportionate increase in those seeing no change. Although Mečiar failed to convince Slovaks of any economic improvements during his tenure, he did manage to give the impression that a Velvet Divorce would have no troublesome impact on the economic sphere and elsewhere. The dwindling of economic pessimists continued throughout the "temporary" government of Moravčík, and then suddenly rose sharply just before the elections – to the great misfortune of Moravčík and all other opponents of Mečiar. The latter's return to the post of prime minister after early elections (October 1993) again caused a fall in the percentage of pessimists and a corresponding increase of neutral positions.

Evaluation of the economic situation in the Czech Republic and in Slovakia has not only been marked by different trends occurring in different time sequences. It also showed significant variation in the degree of optimism or pessimism with which the subject was approached. Since the beginning of 1992, the image of the Czech economy in the eyes of Czechs has, on average, been significantly better than that of the Slovak economy in the eyes of Slovaks. Most (53%) Slovaks feel that the economic situation of Slovakia is deteriorating, while 29% see no change and only 16% are optimistic with respect to future development. In a long term perspective, however, the image of Slovakia's economy with Slovaks has undeniably improved over the years.

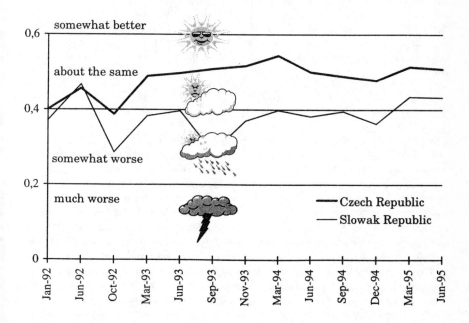

Figure 2 Expected economic development within the next 12 months

Slovak citizens also see the immediate future in a more negative light, although the prevalence of pessimists over optimists is not as great as it was only a short while ago. Not unlike the situation in the Czech Republic, Slovaks expect conditions to remain stable within the coming year, a tendency increased by Vladimir Mečiar's victory at the polls. Czech estimates for 1996 are still more optimistic, although we should note that average expectations in the future development of the Czech economy have remained stable rather than improved. This average rating might be interpreted as indicative of a growing faith in the stability of the Czech economy and subsiding fears of collapse. It is nevertheless true that the average CR citizen initially expected more rewards from the market economy, and would have liked to see the take-off occurring much sooner. Therefore, public evaluation of 1995 must be seen as an expression of the hope that the situation will not get worse, but also as indicative of a general sense of stagnation.

We may thus conclude that in the Czech Republic, the general rating of the economy in the long run shows a prevailing, cautious optimism gradually changing towards a perception of stagnation. In the SR, the development moves rather from a pessimism and feelings of gradual decline to a conviction that things cannot get much worse regarding the Slovak economy, and that worries about further deterioration are futile.

Economic situation of households

The market economy and its system of price regulation has had a profound impact on the perception of social structure. Newly emerging stratification is most evident in opinions on the financial situation of individuals and households. In both republics, there appear to be large numbers of people who see their financial situation as having deteriorated. More precisely, 37% of Czech and 44% of Slovak households reported losses in buying power compared to the previous year. Compared to recent years, the share of people judging their living standards to have dropped is, however, currently declining in both countries. As everywhere, poverty objectively and subjectively hits people with only elementary education, unskilled or semi-skilled employees, and those nearing retirement age.

Among the upper strata of society, the boom engendered by marketization has produced a category of citizens whose financial situation has been improving steadily. Among them, we predictably find a higher percentage of university graduates, entrepreneurs and senior civil servants, who make up the share of "lucky ones" ranging from 16% to 18% in the Czech Republic, and from 9% to 11% in Slovakia. In other words, Slovakia has a smaller share of transition winners, who also require a larger number of compatriots to lose out – a situation reflected in the average ratings of the financial situation of households. In the CR, this evaluation remains just above the "pessimist" bracket, suggesting that, in the eyes of interviewees, the financial situation of Czech and Moravian homes has remained more or less stable. In Slovakia, the average rating of household incomes reflects the perception of a slight, long term decline in the purchasing power of Slovak families.[3] During the time period monitored, evaluation of the economic situation of Slovak families was positive only once. Only in a long term perspective is it possible to discern a slightly optimistic trend, manifest since approximately the second half of 1993 (i.e. after the Slovak Republic became independent).

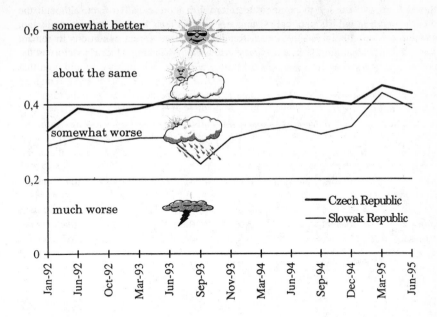

Figure 3 Financial situation of the household

As a result, we may say that SR citizens assess the financial situation of their homes as somewhat better than the economic situation of their country. They thus approach Czechs's average rating of their buying power, who, however, view their own financial situation as poor in comparison to the general state of the Czech economy.

Political culture

Evaluation of the overall economic situation and of individual or household financial conditions following the breakup of the federation thus shows the opinions of Czechs and Slovaks to be gradually converging. This is due mainly to spreading perceptions of stagnation in the CR and simultaneous increases in long term Slovak optimism. Normally, these findings would indicate that Czech dissatisfaction with the management of the economy and with activities of social and political institutions is likely to increase, while Slovak citizens could be expected to show more and more approval with economic and other policies. But the available data on satisfaction with political development in the two countries in no way validate this assumption. Instead, they clearly point in the opposite direction.

In the Czech Republic, credibility of institutions has either remained stable or improved. In 1994, both government and parliament enjoyed much more support than did homologous Slovak institutions: But in the Slovak Republic, trust in institutions has been declining, and especially so in the case of government, whose credibility rate has dropped from 49% to 31% between 1992 and 1994.[4] While CR citizens' satisfaction with democracy and the political system generally showed a 9% decline (from 73% in 1992 to 64% in 1993), in the Slovak Republic this decline was much more dramatic: towards the end of 1993, only 31% were satisfied with the political system, which represents a 32% drop in comparison with 1992.[5]

Table 1 Confidence in Institutions

in percent	Czech Republic				Slovak Republic			
	1991	1992	1993	1994	1991	1992	1993	1994
army	52	42	40	39	31	52	44	47
police	33	36	36	36	35	35	31	31
courts	39	38	32	40	39	37	29	33
public administration	25	27	24	27	28	30	26	26
government	50	48	48	56	23	49	33	31
parliament	32	19	25	32	34	24	20	22
churches	27	33	27	28	21	45	39	45
media	42	40	41	43	47	39	39	38
political parties	22	20	20	24	20	20	13	15

Note: Values 5–7 on a 7–grade scale from 1,0 (= no confidence) to 7,0 (= strong confidence)

Source: Fritz Plasser and Peter Ulram, 'Monitoring Democratic Consolidation: Political Trust and System Support in East-Central-Europe', International Political Science Association, Berlin 1994.

Satisfaction with the political system presupposes a system by and large responsive to citizens' expectations. As regards transparency and opportunities for participation, answers show that both countries' politics are seen in a more or less similar fashion by their citizens, and that such views have also been unaffected by the ČSFR's breakup. The reality of democracy seems to have had a disillusioning impact – three quarters of CR and SR citizens feel that "ordinary people" have no influence on government, that politics is a far too complicated matter for them to be involved in, or that they do not know, on the whole, what it is about.[6]

The huge differences between Czechs and Slovaks with respect to the credibility accorded to the political system of democracy cannot be explained by reference to Slovaks' authoritarian propensities alone. In 1994, single party rule was given preference over pluralism by only 6% in the Czech Republic, and by 20% in the Slovak Republic.[7]

Since the political system and its institutions can be seen as an arena for political actors, that is, one in which they struggle to represent the interests and concerns of their supporters effectively, the behaviour of such actors must count as a decisive factor in the shaping of political culture. If their conflicts are resolved or dealt with in a reasonably civil manner, and if political struggle remains within the bounds of the law, this inevitably has repercussions on the public perception of, and attitude to, the political process. It is in this respect that the Czech and Slovak republics differ most widely. While the behaviour of politicians in the CR does not seem to pose any significant threat to the development of a "civil" society, Slovaks are much more used to dirty politics – a fact reflected in the fearful pride of place they accord to the category of "political intolerance". In the Czech Republic, by contrast, one out of four respondents views racial intolerance as the greatest danger, making it the most important item on the list, narrowly topping "intolerance of opinions".[8] We may say, therefore, that civility of politicians is rather a secondary issue in the Czech Republic, while it represents a major social and political problem in Slovakia. The fact that Slovaks are much more frequently concerned with political extremism than Czechs obviously corroborates this analysis.

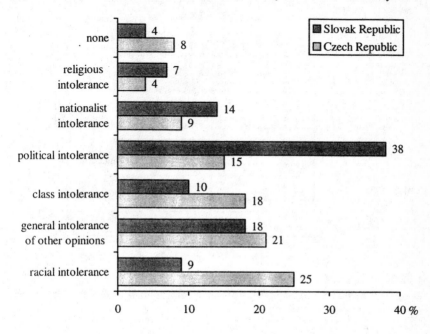

Figure 4 Which type of intolerance do you find the most dangerous?

The issue of intolerance as the rallying cry

A considerable portion of CR and SR citizens are thus aware of the threat which intolerant behaviour poses to society. Many also understand the present situation as an appeal to active interference. Roughly 40% of respondents in both countries declared their readiness to take to the streets and speak out against various forms of intolerance. This remarkable extent of public mobilization is characteristic especially of Slovaks. Their declared intention to join "an immediate action" (e.g. road blockage, occupying a building, etc.) is twice as large as in the Czech Republic.

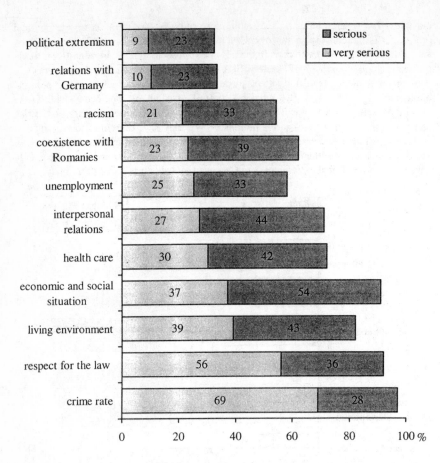

Figure 5 Which of the following constitute, in your opinion, a serious problem for Czech society?

Mobilization of Czech and Slovak citizens is comparatively high[9], but since it is somewhat lacking in practical commitment, it appears to be either a potential or merely a professed willingness to take action. Most people in both countries wish to do something about the problem of intolerance, but few have actually done so. Apart from the fact that many have other, perhaps more pressing problems to address, causes of this disparity between declared intention and action should be sought in deeply ingrained life strategies of caution and conformism, of bowing to necessities, of risking nothing, living peacefully and not standing out to much.[10]

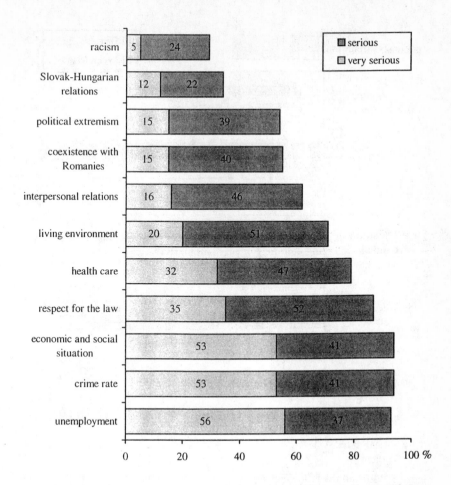

Figure 6 Which of the following constitute, in your opinion, a serious problem for Slovak society ?

Reluctance to participate is also, however, a result of the fact that many do not receive mobilizing impulses or stimuli from various civil associations, initiatives and movements that have created a space in which to realize their protest potential. Although such association do of course exist, communication between them and a larger public seems to be deficient.

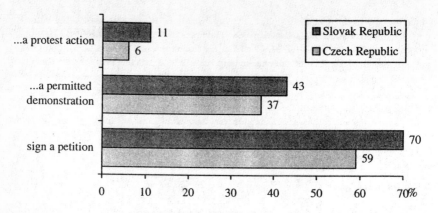

Figure 7 Would you be willing to take action against intolerance and take part in...(respondents agreeing definitely or predominantly)?

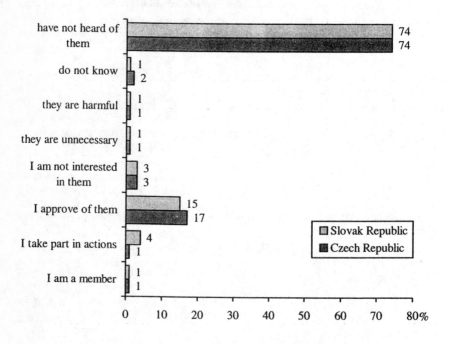

Figure 8 How do you relate voluntary organizations supporting tolerance?

Three quarters of CR and SR citizens are unaware and have never heard of any voluntary organization supporting tolerance on the level of personal relations. Of those interviewed, the few who have heard of such organizations usually do not remember their names. In the Czech Republic, the organizations mentioned most often were the Green Movement (2%), "Nadace O. Havlové-Výbor dobré vùle" (O.Havlová's Foundation of Good Will) (2%), Christian organizations and unions (1%), and charity organizations (1%). "Hnutí obèanské solidarity a tolerance" (Movement for Civic Solidarity and Tolerance – HOST) was identified by only three respondents out of a sample of more than one thousand interviewees. In the SR, the most frequently mentioned are the movement "Human" (1%), Christian organization and unions (1%), and charity organizations (1%). Non-governmental associations supporting tolerance can count on active support of 2% among Czech, and 5% among Slovak citizens. Non-active support reaches levels of 17% and 15% respectively.

Group intolerance

Intolerance may of course be collective as well as individual, and this is especially true in cases of polarized or mutually countervailing forces within society. Tolerance or intolerance between such groups can contribute greatly towards shaping the social and political climate in one way or the other. We cannot categorically assert that only civil relations between groups – such as those commonly found between politicians and journalists, or between mainstream left and right wing parties – have desirable consequences for the political process. But for any general evaluation of a political and social system, the perception of groups and relations between groups as tolerant or intolerant is a crucial indicator.

In both countries, respondents generally shared the view that out of eight selected structurally antagonistic groups, the most cordial and constructive relations prevailed between employees and employers, as well as between youths and the elderly. In the SR, greater emphasis was placed on the tolerance of employees and older people. The overall impression was that of prevailing tolerance between the selected groups in the SR, and of predominantly intolerant relations between the same groups in the CR. This should not be taken to mean, however, that the social atmosphere in Slovakia is less charged or conflict-ridden. Slovak inter-group relations seem to be more asymmetric, that is, marked by intolerance coming mainly from one side without being reciprocated proportionally. Answers show that the poor are thought to be magnanimous towards the rich, while the latter are seen as callous towards the poor. A similar situation is found in the case of employment relationships and those between unions and government. Employees and unions, it is thought, face much stronger intolerance from their opponents than vice versa. Asymmetric intolerance in Slovakia is also prominent in the relationships between groups instrumental in the process of social reconciliation. The role of tame sufferers is typically attributed to poor employees or union members, while powerful employers as well as government are cast in that of arrogant, hazardous gamblers. As yet, it is neither clear how much longer this asymmetry can continue, nor who will eventually have to modify their behaviour.

The perception that the fiercest antagonism reigns between politicians and journalists revealingly describes the current political turmoil in Slovakia. In the Czech Republic, the interest antagonism most frequently characterized as intolerant is that between trade unions and government. But it is important to note that interviewees saw intolerance as almost evenly distributed between either camp. Here, as in other such binary sets, uncompromising stances are seen to be symmetrically distributed. One important implication of this is that overly partial or prejudiced resentment, and explosive tensions resulting from them, are less likely to accumulate in the Czech Republic. Intolerance in the Czech Republic tends not to be concentrated to the same degree. Perhaps also in connection with upcoming elections, conflict-laden politics have become a much more generalized feature of the Czech Republic's political system.

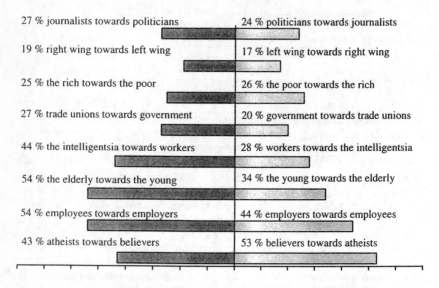

27 % journalists towards politicians — 24 % politicians towards journalists

19 % right wing towards left wing — 17 % left wing towards right wing

25 % the rich towards the poor — 26 % the poor towards the rich

27 % trade unions towards government — 20 % government towards trade unions

44 % the intelligentsia towards workers — 28 % workers towards the intelligentsia

54 % the elderly towards the young — 34 % the young towards the elderly

54 % employees towards employers — 44 % employers towards employees

43 % atheists towards believers — 53 % believers towards atheists

Figure 9 Czech Republic, tolerance show

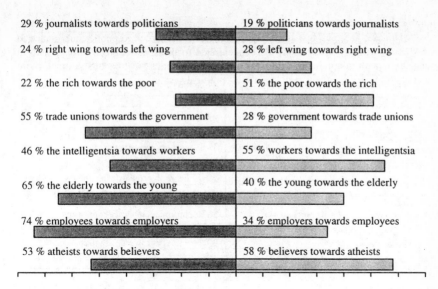

Figure 10 Slovak Republic, tolerance show

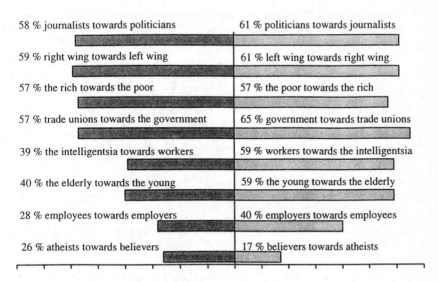

Figure 11 Czech Republic, intolerance show

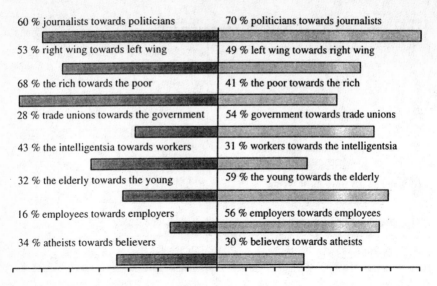

Figure 12 Slovak Republic, intolerance show

Notes

1   The following findings are based on data collected as part of a continuous research project entitled 'Consumer Barometer' and carried out quarterly by GfK Praha, using a representative sample of 1,000 CR and 500 SR citizens. The first survey took place in January 1992, the last at the beginning of June 1995.

2   The average value (index) has been calculated as follows: 5–degree scale (much better, rather better, about the same, rather worse, much worse, "undecided" was not included), each question was weighed on a scale of 0 to 4. The weighed average was subsequently divided by five. In this way the resulting figure was transposed into the interval between 0 and 1.

3   See Pavol Frič, 'Consumer Barometer CR+SR', GfK Praha, June 1995, p. 6.

4   Data are taken from periodic surveys by Fessel+GfK, Paul Lazarsfeld-Gesellschaft, and NDB-Neue Demokratien Barometer. See Fritz Plasser and Peter Ulram, 'Monitoring Democratic Consolidation: Political Trust and System Support in East-Central-Europe', International Political Science Association, Berlin 1994.

5   ibid.

6   Data are taken from periodic surveys by Fessel+GfK, Paul Lazarsfeld-Gesellschaft and NDB-Neue Demokratien Barometer. They cover the period from 1990 to 1993. See Stanislav Zahradníček, Neue Demokratien Barometer, GfK Praha, Praha 1993, p. 7.

7   See: Fritz Plasser and Peter Ulram, op. cit.

8   These and subsequent data concerning the issue of intolerance are taken from the Gfk Praha survey carried out during the first half of 1995 on a sample of 1020 CR and 472 SR citizens. See Pavol Frič, 'The issue of tolerance and intolerance in CR and SR', GfK Praha, Prague 1995.

9   In Britain, 14% of respondents in 1991 were willing to participate in protest rallies in cases of injustice. See Jowell, R., Brock, L., Prior, G. and Taylor, B. (eds.), British Social Attitudes, 9th Report 1992/93 Edition. Cambridge University Press.

10   See Pavol Frič, 'The Life Style', GfK Praha, Praha, 1995.

14 Political Culture in the Czech Republic

Emanel Pecka

In a large number of books and articles on the subject, we read that "political culture" defies precise definition. At the same time, it is agreed that the concept may be useful for descriptions of political behaviour generally, rather than being limited to the narrow field of civic conduct; it includes everyday behaviour, distinctive customs, skills, and attitudes individuals adopt as part of a shared experience of their political system. If we are to understand contemporary Czech political culture comprehensively, we should therefore seek to identify its roots and then describe its main features.

Any attempt to explain the development of political culture in the Czech Republic must begin by examining the country's democratic traditions. The creation of a democratic state on our territory was to a large extent the result of long efforts on the part of Czech politicians drawing on ideas of nineteenth century thinkers: J. Dobrovsky, J. Jungmann, J. Kollár, and, perhaps most of all, F. Palacky. Thomas Masaryk, for one, often quoted the latter as saying that states evolve in the spirit in which they are born. The new state introduced parliamentary democracy and human rights to Central Europe, and the Czechoslovak public grounded on this a patriotic pride which laid the foundations for Czech nationalism.

Within twenty years (1918–1938), democratic traditions had become the dominant components of our nations' political culture. People from across the political spectrum carried on these traditions throughout World War II, and even during the totalitarian period. Although attitudes to democracy were far from homogeneous, there were some conceptions or basic political and cultural orientations shared from the beginning, and by nearly all important social groups. Three important books published during the 1920s and dealing with the relationship between culture and politics might give an indication of this broad consensus, as well as of the lively debate surrounding it: Edvard Beneš's "The Nature of Political Parties" (1920), Vilém Mathesius's "Cultural Activism" (1925) and Ferdinand Peroutka's "What are we?" (1924), a belated reply to the Czech sociologist Emanuel Chalupny and his book on the "Character of a nation" (1907). The sociologist Edvard Beneš, then Minister of Foreign Affairs, thought that Czech political culture was deficient and saw the way to an improvement in political education towards democracy. Guiding this process of instruction was, in his eyes, the main task of political leaders, who should devise long term strategies of shaping democratic values. The masses, Beneš asserted, lacked direction and were ready to assume the values of a governing elite. In short, Beneš may be taken to represent a line of political thought emphasizing the role and authority of national leaders. In public debates during the years

1924 and 1925 especially, several academics and journalists dealt with the question of "national character". The journalist Ferdinand Peroutka, for instance, espoused the view that in the case of Czechs, centuries of foreign domination were responsible for what he considered a political culture of fatalism. Literary scholar Vilém Mathesius was more sanguine about the possibility of effecting change, claiming that, with fair and competent leaders, his nation could promote a culture of activism. Sociologist Emanuel Chalupny stressed the importance of reformatory traditions in identifying a culture of anticipation or millenary hope as a defining feature of Czech national character.

Whatever the eccentricities of this debate on "national character" from today's point of view, we may be able to glean from it some concepts for defining aspects of Czech political culture: patriotism and nationalism, a high degree of respect for national leaders, as well as elements of fatalism, cultural activism and hope. One or another of these aspects may have dominated at one time or another, and it is becoming increasingly clear that these features, as well as the public debates of the interwar period, are still relevant for an understanding of contemporary political culture.

Since the Czech Republic is in the process of creating a new post-totalitarian set of social norms and values, its political culture could be seen as a transient one. Since November 1989, the problem of how to manage democratic tradition, as well as its corollary of what to do with the authoritarian past, has been with us. Another equally important and troubling question has yet to be answered: was this past as a whole totalitarian? After all, more than two generations lived and took part in the formation of totalitarian society, and the key elements of that order cannot have failed to become part of their mindset. During 42 years of communism, conformism was a precondition for survival and stigmatized everyone living in Czechoslovakia, irrespective of his or her political views. But at the same time, we need to be aware that the Prague Spring or the process of civic revival generally cannot have developed out of a void. Much in the daily lives of Czechs prior to 1989 likewise testifies to the implausibility of the assumption of a thoroughly totalitarian political culture.

For our research of political culture carried out between 1968 and 1970, we used Almond and Verba's three categories of cognitive, affective, and evaluational orientation (Pecka, Pecka and Ungr 1992). As for cognitive orientations of Czech citizens, we found that in 1968 people were primarily concerned with positioning themselves in society, and with defining their political identity and aspirations. "Who are we?", "What is our social idea?" and "What do we expect?" – such were the questions found to be of primary relevance. In March 1968 (following the abdication of president Antonin Novotny), 88% of the population expected some form of progress of both socialism and democracy. In June and July of the same year (after the so called "Appeal of the Two Thousand words" and after pointedly menacing Warsaw Pact maneuvres had come to an end), 86% of respondents declared to favour an advancement of socialism, and only five percent opted for capitalism. According to the results of a survey conducted by the Institute for Public Opinion, a majority of people thus expected that reforms of the political system would lead to democratic socialism. But what exactly, in their eyes, did that coinage denote?

No doubt, civic activism was fuelled by the events of 1968, and people had clear expectations regarding standards and conditions of living, as well as the changes in governance and leadership necessary to fulfill those demands. With respect to economic reform, the results of the survey mentioned above indicate that more than one third of respondents were bracing for the possibility of a further decline in living standards within the short term range, while hoping for marked improvements within a period of five years. There prevailed, in other words, a remarkable degree of civic realism. Although living standards were understandably a central element in the hopes for thoroughgoing reform, greater attention was paid, on the whole, to its political aspects. Not least, the events of 1968 brought with them an unprecedented degree of politicization.

In May 1968, a majority of respondents expressed confidence in mass media. When confronted with the question "Do mass media report on problems with insufficient competence?", 80% thought that the media informed truthfully and candidly. Trust in the work of journalists and a feeling of loyalty towards them were widespread at the time. Workers in Prague and Ostrava, for instance, founded a "works council for the protection of the press". 86% of respondents expressed their agreement with the decision to remove censorship. The journalist Helena Klimova characterized the relationship between the public and the media in the following way: "Civic consciousness suffered from schizophrenia during the last twenty years. It was engaged in two parallel monologues: the monologue of official public opinion and that of genuine but private opinion. This schizophrenia was healed during the Spring and Summer of 1968, and by the time the tragic weeks of August came round, 98% of inhabitants had developed or renewed their confidence in journalists".

The events of 1968, and especially military intervention, had immediate repercussions on citizens' affective political orientation. This was expressed in various forms of public protest, on posters and in slogans, and began to circulate in the form of jokes and rumours. All these served as forums for the expression of solidarity, and are ample evidence of the general climate of activism even in the face of a situation more than conducive to fatalist resignation.

Theoretical contributions are perhaps the more valuable of those belonging in the category of "evaluational orientations". The eminent Czech philosopher Karel Kosik published several articles on "Our Current Crisis" (the title of a book by Thomas Masaryk published in the 1890s) just before the invasion. He described the political crisis as not just a struggle for existence but, in the spirit of Masaryk, as a negotiation over the meaning of national and human existence. "Have we fallen to the level of an anonymous mass for whom conscience, human dignity, the sense of truth and justice, honor, decency, and courage are unnecessary ballast, or are we capable of remembering and solving all our economic, political and other questions in harmony with the claims of human and national existence?" Kosik's objectives were the "political equality of rights and fullness of rights derived from the principles of socialism and humanism." His writings reflected a general mood of optimism and of the recuperation of Czech and Slovak pasts, and were indicative of the ongoing hope for a more satisfactory solution of the old "Czech question". The traditions of democracy and patriotism, which had been submerged under Nazism and Stalinism, surfaced once again. Kosik's thought is also part and evidence of a continuing culture of hope and activism, though we must remember that he was writing just before the crushing of the Prague Spring (Skilling 1976).

A symposium on "The Czech Question" held in March 1969 saw historians, writers, philosophers and literary scientists deliberate on the democratic ideas and actions of Thomas Masaryk. The idea was to revive and re-legitimize them, to draw on the authority of a great name rooted in Czech collective memory in order to integrate them anew into contemporary Czech political culture. As Gordon Skilling put it,

This revival of past traditions lends support to the view that the movement for reform in 1968 was a product of certain crucial elements in traditional political culture, such as the humanist values of Masaryk, democracy, freedom of expression, Slovak independence, and Czech-Slovak solidarity (...). Yet 1968 was more than a renaissance of the past. It represented an advance toward a more humane social order embodying in some degree the tradition of Masaryk and Marx, both critically reinterpreted, and synthesizing the basic ideas of democracy and socialism in a manner not only new to Czechoslovak history but unduplicated elsewhere in the world. Yet there were negative aspects of the Czech and Slovak past which could not be ignored even in the heady atmosphere of the Prague Spring, and which were bound to loom larger in the depressing post-occupation times. It was recognized, in a characteristic spirit of self-criticism, that Czechs and Slovaks had often failed to find a satisfactory answer to their problems and that this was sometimes due, not only to external factors, but to the lack of will and the failings of the Czech and Slovak people and their leaders.

The years 1968–1970 not only saw a revival of historical awareness, but also the invigoration of old traditions in political culture in a situation of national duress. It was during these revolutionary years that these traditions took their modern shape without being transformed beyond recognition: Anticipation as a vision of socialist democracy; veneration of national leaders as an anchorage in the search for protection against Soviet coercion; activism as a way of addressing the Czech question in a new form; patriotism and nationalism as a strategy of unifying Czechs against military threat and violence; but also fatalism resulting from defeat.

These elements, in conclusion, could still be said to make up contemporary Czech political culture, though with a somewhat weaker framework of national unity, which has partly been replaced by the pluralism of political parties. New divisions are emerging, particularly the dualism of "left" and "right", which constitutes an increasingly important, though somewhat shifting, divide. Differentiations of this kind are most visible in their alternative normative references, in a different interpretation of universalist and achievement values, and in different conceptions of the relationship between individual aspirations and the requirements of society. The right-left division will, of course, in the long run need to be integrated into existing patterns of political culture. Only then can the new pluralist democracy be provided with a bracket or basic consensus needed for solving the problems of a rapidly changing society.

References

Almond, G.A. and Verba, S. (1965), *The Civic Culture*, Little Brown & Co., Boston.
Beneš, E. (1920), *Pohava politického stranictví*, Prague.
Chalupný, E. (1907), *Národní povaha česká*, Prague.
Klímová, H. (1969), 'Co si myslíme u nás doma', *Listy*, 6.
Masaryk, T.G. (1948), *Česká otázka, Naše nynější krize*, Prague.
Mathesius, V. (1925), *Kulturní aktivismus*, Prague.
Pecka, E, Pecka, J. and Ungr, V. (1992), *Každodennost konce 60. let*, USD Akademie věd České republiky, Prague.
Peroutka, F. (1924), *Jací jseme*, Prague.
Skilling, H.G. (1976), *Czechoslovakia's Interrupted Revolution*, Princeton.

SOME CONTRASTS FOR COMPARISON

15 Aspects of Political Culture in Austria

Karin Liebhart

Some methodological considerations

In methodological terms, "political culture" comprises very heterogenous approaches of analysis. Studies on political culture in Austria are mainly oriented towards a typology of the political system and its essential or characteristic traits (cf. Hanisch 1995, Pelinka 1995, Plasser 1994, Plasser and Ulram 1991, Ulram 1990, Gerlich 1989). These studies deal above all with an area described as "civic culture" by Almond/Verba (1963), and whose scope includes input, output, and system orientations, attitudes and behaviour patterns directly influencing the political system or its subsystems (parties, lobbies, elites, bureaucracy, etc.), as well as factors immediately acting on the political system (such as mass media). Such long term typologies of Austria's political system have been attempted by, among others, Ulram (1990) through political field analysis (attitudes and expectations regarding the political system), by Hanisch (1984, 1991, 1994) and Pelinka (1985, 1995) through the inclusion of historical continuities, by Gerlich (1989, 1991) through an analysis of political taboos, i.e. of attitudes of denial or deliberate marginalization (cf. Dachs et al (ed.). 1991, 419–497) and by Plasser and Ulram (1991) in their comparison of Austrian, Swiss and German political cultures.

The methodological connection of such studies with, for instance, research on stereotypes is expressly stated in the works of Almond and Verba (1963, 1980) as one preliminarily defining blueprints of national or social "characters" for the purpose of studying political culture (Thompson, Ellis and Wildavsky 1990, 219). This approach espouses a – Durkheimian and Parsonian – functionalist understanding of politics and society, in which norms and values feature as durable social facts, as "the glue or cement of a society" (Gibbins 1989/90, 4) determining and explaining political behaviour beyond the range of incidental short term variation.

However, international comparison shows this approach to be only one of several possible paradigms for the study of political culture. Owing to the difficulty (increasingly encountered during the 1980s) of explaining political change, and in order to explain the growing distance between politics and "the public" as structurally inherent in the political system and its associated spheres, international debate has increasingly taken recourse to the concept of "lifestyles" – a trend echoed in Austria by Fessel+GFK's lifestyle studies of 1987 and 1992, as well as in the research of Heinrich (1990, 1991), albeit in different ways. The term "(social) lifestyle" has so far mainly been applied in the sociology of culture following Pierre Bourdieu (Mörth and Fröhlich

1994). As a concept in political culture research, it has its origin in two separate strands of social theory which, in turn, reflect recent social and political developments.

On the one hand, there is a need to try to explain the appearance of fringe ideologies and groups in Western political systems, which have either emerged spontaneously or are now surfacing after having been marginalized for some time, such as ecological or New Right movements. Studies of political culture have reacted to this proliferation of topics by going beyond the concept of civic culture, and by turning to other spheres of everyday life as well as to the value concepts and attitudes shaping them. Conditions of milieu provide the frameworks for patterns of culture and politics, and may be seen as mainsprings of such changes in patterns as cannot be explained by reference to politics alone. The results of lifestyle analyses have led to the assumption of a "post-materialist" trend within Western societies, which is said to provide the background for such changes in the political system (cf. Gibbins 1989/90, 8ff.). This links the studies in question to the postmodernist debate. Daniel Bell (1973) uses the term "postmodernism" to denote the consequences of transition to a service economy for the structure of society: an economic trend which both engenders new social groupings (an economically relatively independent urban middle class) and marginalizes existing groups by way of rationalization processes. In keeping with their changing interests, these newly formed groups consequently enter the political sphere as protest potentials of various and diverse orientation.

> Inglehart's (...) predictions that new forms of political expression would emerge, involving new and somewhat unconventional political groupings, came to fruition in the form of alliances around ethnicity, environmentalism and peace; new movements such as feminism and new trade unions. Indeed new symbolic and lifestyle politics emerged in the form of Green politics and the concentration upon images and optional lifestyles in conventional party performance and literature. These developments have in turn supported the claim that there was a gradual move from 'old' to 'new' politics in Western Europe, with the new politics being an expression of postmaterialistic values (...)" (Gibbins 1989/90, 9)

On the other hand, we need to raise the question whether an ethnocentric perspective influenced by Almond and Verba's concept of political culture as "civic culture" and modelled on the structures of Western political systems – one "premised upon an ethnocentric, Anglo-American confidence in liberal democracy" (Gibbins 1989/90, 8) – can or should be applied in research on political cultures which have developed in different contexts and under different conditions. In response to such queries, a typology of lifestyles has been developed which approaches different societies anthropologically, that is, partly in their own terms – a trend in current sociological debates aiming at the relativization of self-images inherent in methodologies. This was further elaborated by Douglas and Wildavsky (1983) to allow for comparisons of political systems on the basis of hierarchy structures within groups. They suggested a typology of lifestyles within a four field matrix (hierarchic to egalitarian, individualist to apathetic). Predominant orientations of societies and political cultures could then be evaluated on the basis of different field profiles. This typology – which in part also served as an explanatory model for Fessel+GFK's lifestyle studies of 1987 and 1992 (though merely as a typology of attitudes) – was further developed to embrace a cultural theory that allows for typifying descriptions of values and attitudes (cultural biases) within their social and political contexts and as behaviour patterns of political cultures. Thompson, Ellis and Wildavsky (1990) have tried to describe characteristics of political cultures on the basis of such a framework from an internationally comparative perspective. In Austria, Heinrich (1990) used the same approach in his effort to reformulate the current debate on "Central Europe" with particular attention given to its rootedness in social structures, and in his comparative analysis of the political cultures of Warsaw and Vienna (Heinrich and Wiatr 1991). This approach permits the inclusion of patterns of symbolic action and interpretation, and of the kinds of social relations they allude to or represent, as constituents of processes of political identification.

By forging a link with the concept of political culture, the ethnomethodological approach allows us to relate descriptive analysis (as applied by, among others, Lynn Hunt, 1984, in her account of changing perceptions of political culture during the French Revolution) to evaluative statements occurring within this political culture. Similarly, Patzelt (1987) has tried to fuse "politological analyses on the micro and macro levels" by adapting Garfinkel's ethnomedological theory (1984) to the needs of political culture research. "The creation and continuation of everyday political culture should be seen as a social construct of self-evident political givens shared by all members of an ethnic group who create and utilize them (...)."[1] This refers to both everyday actions and "theories" or – however casual – opinions expressed with respect to these realities. "In ethnomethodology, the scenic practices of utilization, construction or deconstruction of self-evident political givens (...) should be isolated."[2] Thus, political rituals that are shared by a particular group and constitute or relate to these givens, such as adherence to a nation, need to be traced.

The national entelechies of different political cultures may serve as a defining characteristic of such political groups, as their (auto-)reflexivity and its expression are in turn part of the self-evident givens, for instance by being seen or interpreted as "typically Austrian" or "typically German". Constructions of national identity or identities as forms of such self-evident political givens obviously belong in this domain, and may also serve as starting-points for their reconstruction.

Political myths, rituals and symbols are forms of "scenic practices" (cf. Edelman 1964, Hunt 1984, Voigt 1989); they can also be conceptualized under the heading of collective memory (Halbwachs 1950) as "stagings" of events. Ethnomethodological studies of particular forms of "staging", as well as comprehensive ones of political cultures as entities or symbolic systems, have been available for some time, e.g. on the construction of documents (e.g. files, laws, political programmes etc.) and on the representation of reality by the media. The latter, of course, play a special part in politics through their "staging" of politics as well as by conveying and (by implication) interpreting "significant events".

The term "culture" is used here with reference to Clifford Geertz, who suggests integrating definitions of culture frequently resorted to in sociology and cultural studies:

> [C]ulture is best seen not as complexes of concrete behavior patterns – customs, usages, traditions, habit clusters – as has, by and large, been the case up to now, but as a set of control mechanisms – plans, recipes, rules, instructions (what computer engineers call 'programs') for the governing of behavior (Geertz 1973, 44).

213

Similarly, Frank Robert Vivelo defines culture as a system of rules or principles for correct behaviour which comprises standards for behaviour but is not behaviour itself – analogous to a linguistic grammar, which can best be seen as a system of rules or principles for correct speech (1988, 55). This understanding of culture may be brought into line with both general sociological conceptions of rules and social roles, and ethnomethodological or similar reconstructive approaches to political culture.

The development of political culture and its context in Austria

In most of the above mentioned studies, the political culture of Austria's Second Republic (1945 to the present) is singled out as a special case within the context of Western Europe (Pelinka 1994). In so far as it qualifies as a liberal democracy, the political system is characterized by competition of political parties; however, it is also marked by a conspicuous – "typically Austrian" – degree of cooperation between the two major parties and by a strong element of corporatism.

For a long time – in fact, until the mid 1980s – this meant an extraordinary concentration of political power, since most Austrians voted for one of the two major parties SPÖ (Socialist Party of Austria) or ÖVP (Austrian People's Party). (The country was governed by so-called Grand Coalitions in 1945, and from 1947 to 1966. Between 1966-1983, single-party governments were in power. During those four decades, other parties generally played only a minor role.)

The resulting pattern of concordance and cooperation, as well as the virtual smothering of political competition that came with it, was caused by a reorientation within the political elites of both parties after 1945, a change of heart which is said to have been the result of crucial historical experiences (cf. Mattl 1995, 38f.), specifically

– the civil war between Social-Democrats and conservative "Christian-Socialists" in 1934, leading to the collapse of democracy and an Austro-fascist regime;
– the annexation ("Anschluß") of Austria by the German *Reich* in 1938 and the experience of National-Socialism.

The compromise between the elites of the two major parties after 1945 has been seen as an antithesis, or simply a reaction, to the breakdown of Austria's First Republic. This compromise shrouded the whole of society in a political culture of concordance (Plasser and Ulram 1991). Proportional segmentation of important subsystems according to the political strength of SPÖ and ÖVP corresponded to an exceptionally high degree of political organisation within society. This system also continues to exert considerable influence on the economy, *Sozialpartnerschaft* or "social partnership" forming part of Austria's often told post–1945 "success story".

Paralleling close cooperation of political elites, loyal party supporters formed the basis of Austria's political system until the mid 1980s. However, these traditional loyalties have long begun to erode. In 1986, the Austrian Freedomite Party (FPÖ) doubled its votes and a new type of party, the Green Alternative, entered Parliament. In 1993, yet another new party, the Liberal Forum (a liberal breakaway from the FPÖ) made its way into Parliament. Not surprisingly, these newcomers have been challenging the homogenous structures of organized interests, which are institutionalized in the framework of social partnership.

214

At the same time, electoral participation and party membership declined, both of which had been extremely high until the early 1980s. In 1975, 93 percent of votes went to either SPÖ or ÖVP, and participation in national parliamentary election until then regularly exceeded 90%. In 1990, only 75% of Austrians favoured one of the major parties, while electoral turnout fell below 90%. Young people, especially, abstain from voting (Pelinka 1994).

What are – in the context of many changes affecting Austrian society – the reasons for these marked shifts in the country's political culture?

- Social mobility, drastic improvements in educational standards, and secularization – which has reduced the power and influence of Catholic Church –, make the boundaries between political parties more permeable;
- electoral core groups (such as small farmers and industrial workers) of the major parties have all but vanished;
- subcultures – such as that of the working class – have been integrated into the political and cultural mainstream;
- the growing complexity of issues makes clear and unidirectional orientation more difficult;
- public interest in political matters has increased, while simultaneously the distance between established political organisations and institutions on the one hand, and their clienteles on the other, has grown. The same may be said of the relationship between voters and political elites of various parties.

Pelinka (1994) has therefore argued that Austria's political culture will generally become more "Western", and that it will adapt to so-called normal standards of older democracies, such as
- a deconcentration of the party system;
- an erosion of "social partnership" and its corporatist structures;
- increasing deregulation and an end to the state's heavy influence on, and interference with, the economy;
- increasing political competition;
- new forms of political participation, e.g. referenda on issues such as the use of nuclear energy or membership in the European Union;
- an upsurge of populism.

Taken together, these trends illustrate the dynamics and pace of political and cultural change since the 1970s (Plasser 1994, 69). This has accelerated since the late 1980s (cf. Dachs 1995, 290ff., who has designated the year 1986 as *Wendejahr* or "watershed") and especially since the Austrian governement's move toward joining the European Union. The breakup of the "Grand Coalition" between Social Democrats and Conservatives in October 1995 is a further symptom of these developments.

Political taboos and ambiguities in the political culture of Austria's Second Republic

So far, descriptions and analyses of Austrian political culture and its context have focused on attitudes and behaviour patterns which come to light in so-called "expressible things", that is, which can readily be made explicit. Unconscious social structures – those which cannot be verbalized – by necessity escape the devices of opinion surveys (cf. Reiterer 1988, VII; Gerlich 1989, 11). But if culture (and political culture) can – as we have seen – be interpreted as a set of control mechanisms and rules for appropriate behaviour (Geertz 1973, 44; Vivelo 1988, 55), an ethnomethodological approach may be helpful in reconstructing hidden dimensions of political culture in the Second Republic.

In the case of Austria, collective historical experience, which shapes political culture and provide the framework for political discourse, may be characterized as pervasively ambigous: its political culture has always been seen as one of ambivalences. Although the founding of Austria in 1945 was in part conceived as an antithesis to German National Socialism, the political culture of the Second Republic has been marked by personal and organisational continuities with that period in many spheres of society (politics, science, arts, administration, and bureaucracy; cf. Kerschbaumer and Müller 1992, Rathkolb 1991).

This tension underlying Austrian political culture have resulted in political taboos – as they have been named by Peter Gerlich (1989) – or "negative aspects" of political culture. Social and political taboos, however, may make it impossible to address important issues, or at least constrain discourses in which they can be dealt with, since a political culture's taboos and "dark sides" determine what constitutes, or does not constitute, a political question (Gerlich 1991, 463). A kind of collective repression concerning Austria's share in National Socialism and its atrocities has been one of the main sources of ambiguities perplexing contemporary Austrian society. This in itself has been a taboo for a long time, or, to be more precise, until 1986, when the revealingly entitled "Waldheim affair" lifted the lid.

Within Austrian society, a tradition of representing the entire country as a victim of National Socialism took hold, thereby denying that many Austrians had condoned and played an active part in it. Over the years, this amalgamation of self-pity and collective repression contributed considerably to the political and cultural make-up of Austria's Second Republic. Repression and downright "deletion" of embarrassing political events from the official Austrian memory became part of popular political culture. This process was part and parcel of Austrian everyday political consciousness, both for the public at large and for political leaders. For political elites in particular, taboos resulting from the denial of important elements of the political past offered certain "functional" advantages. They also served politicians from both major camps in the tactic of "political hostage-taking" (Pelinka 1995, Ringturmgespräche 1995). In their (tacit) agreement not to raise troublesome issues relating to the National-Socialist period, representatives of SPÖ and ÖVP tried to achieve political integration by renouncing their integrity.

This assertion is borne out by various political scandals accompanying the "pre-Waldheim era" as well as by monuments erected in honour of Austrian war criminals and by the practice of naming buildings after National Socialist celebrities (eg. the Moritz Etzold hall in Wels, Upper Austria). Even after 1986, controversies such as those surrounding the attendance and hence legitimation of *Kameradschaftsbünde* and their public events (paramilitary "comradeship" groups of self-described true patriots, who nevertheless fought for Nazi Germany rather than Austria during World War II) by members of parliament are ample evidence of the weight of continuity.

216

Largely half-hearted and inconsistently implemented de-nazification measures (cf. Stiefel 1981, Meissl et al. 1986, Sternfeld 1990), which were soon to be followed by generous amnesties for the "less incriminated", fit the picture of Austria as Hitler's first victim. Taken together, these dilatory tactics proved highly opportune and politically auspicious after 1945. The majority among incriminated elites in science, art, journalism, law and public administration retained their positions and were largely spared any de-nazification measures whatsoever (cf. Kaindl-Widhalm 1990, 31).

Persistent attempts to integrate the persuasive myth of a victimized Austria into the nation's post-war self-image – an interpretation relying considerably on official documents, such as the Moscow Declaration of 1 November 1943 – have been characteristic of Austria's fundamentally ambivalent political culture (cf. Gerlich 1991, 463, who analyzes this phenomenon as *Doppelbödigkeit*, or ambiguity).

To the outside world, Austria was depicted as overrun by Nazi Germany and subsequently liberated by allied forces. It vehemently distanced itself, in its foreign policy, from the National Socialist past it had shared with Germany. In contrast to the First Republic, the Second construed itself as resolutely anti-German (and, by implication, anti-Nazi). At home, however, the two major political camps, SPÖ and ÖVP, pursued parallel strategies of propagating a clean slate image while re-integrating former Nazi leaders and collaborators into politics. These had been barred from the polls in the early days of the Second Republic, but soon began to represent a coveted reservoir of potential votes.

Contrary to the self-styled image of Austrian government during the early post-war years, representatives of the Second Republic also boycotted the return of emigré Jews (Reinprecht 1992) and the restitution of "Aryanized" property (cf. Knight 1988). Apartments, businesses and other property confiscated in the name of "Aryanization" were rarely returned to their original owners. Reparation measures were carried out with great hesitation in order to avoid drawing attention to the fact that many Austrians helped implement genocidal racial policies; for certain groups of victims (such as homosexuals) reparation payments still have not been made (cf. Bailer 1993 and the present debate on the "Fund for the Victims of National Socialism" which disburses compensation payments as an act of grace rather than one of meeting an obligation or justified claim).

Shifting responsibility for National Socialism onto Germany involved a "deletion" of a considerable portion of the nation's history. This in its turn produced a "political culture of unconsciousness" (Ziegler and Kannonier-Finster 1993, 39), i.e. one marked by the absence of serious discussion of awkward topics tabooed by the political elite's "silent complicity" (Ziegler and Kannonier-Finster 1993, 244). The experience of National Socialism as a moral imperative and normative basis for self-reflection never really took hold in Austrian political awareness.

As mentioned above, former Federal President Kurt Waldheim's attempt to downplay his National Socialist past, as well as his statement during the presidential election campaign of 1986 that he had "only done his duty", illustrate this deeply rooted Austrian attitude only too well.

> Waldheim's none-too-convincing explanation that he had only done his duty made the notion that Austria had been Hitler's first victim seem implausible, and branded this attitude as an opportune half-truth of the *Staatsvertrag*, or State Treaty, generation. As an unintended side effect, Waldheim's dictum of "fulfilling

duty" towards the German *Wehrmacht* attracted attention to the question of Austrians share in the crimes of World War II. Although Waldheim's key statement – and as such it was generally interpreted – rang truer (to the attitudes of the war-time generation) than any other explanation, it also ran counter to *raison d'état* and to fundamental doctrines of the Second Republic.[3]

The Federal President, Kurt Waldheim, became a symbol for widespread dishonesty and the absence of a candid public debate concerning the Nazi period. According to Robert Schediwy, however, almost all actors on the stage were themselves culpable in some way or other (1992, 54). Even Waldheim's outspoken critics proved to have been taken in by myths distorting the past; they had turned a blind eye on the cynical or opportunistic stance of other politicians and had shown indifference in other, similar cases.

In 1975, even the so-called Kreisky-Peter-Wiesenthal affair did not result in a debate infringing on the taboo topic of National Socialism in Austria. (Bruno Kreisky, then Austrian Federal Chancellor, came to the defence of coalition partner and leader of the Austrian Freedomite Party (FPÖ), Friedrich Peter, against – substantiated – accusations that he had voluntarily served as *Obersturmbannführer* in an SS batallion implicated in mass murder. Although Kreisky had himself been a victim of Nazi persecution, he accused the head of Vienna's Jewish Documentation Center, Simon Wiesenthal, of using "mafia methods".) Only the handshake between Austrian Defence Minister, Friedhelm Frischenschlager, and SS war criminal Walter Reder – arriving at the airport after having been pardoned by the Italian government – finally succeeded in causing significant reverberations both nationally and internationally (cf. Gehler and Sickinger 1995, 678).

Josef Seiter (1995, 702) confirms that Austrians' rapport with their more recent history is "still problematic". As a case in point, official monuments still reflect what I have been calling an ambivalent consciousness, and the attempt to "delete" embarrassing historical events by omission is readily apparent. Historical evidence for this is provided by historian Heidemarie Uhl concerning the province of Styria (1994, 111–196; cf. also Gärtner/Rosenberger 1991)[4]. A similarly unease with respect to national history and its implications for political culture can be detected in the way Austrians relate to political symbols. Thus, the shattered chains around the claws of the Austrian Eagle, which are supposed to symbolize liberation from National Socialism, have lately been a source of considerable controversy.

Today, the fact that Austria's National Socialist past is in part no longer concealed allows for official acknowledgment of its responsibility for the Holocaust by political leaders such as chancellor Vranitzky in 1991 and 1993, and federal president Klestil in 1994. This, however, has been a double-edged sword in more than one respect.

National Socialist propaganda has again become a possibility in Austria; openly antisemitic and xenophobic views are successfully clamouring for political and social acceptance, thereby revealing antisemitic and xenophobic structures at the base of Austrian political culture (Wodak 1990). Xenophobia in politics is particularly salient in asylum policies, or in immigration legislation. Increasing hostility towards minorities like Roma and Sinti, or towards other groups designated as foreigners or outsiders, has made inroads on political discourse as well as political action: acts of terrorism and violence have been directed against such groups, and have become symbolically anchored in public consciousness by the name of Oberwart, Province of Burgenland (cf.

National Socialist propaganda has again become a possibility in Austria; openly antisemitic and xenophobic views are successfully clamouring for political and social acceptance, thereby revealing antisemitic and xenophobic structures at the base of Austrian political culture (Wodak 1990). Xenophobia in politics is particularly salient in asylum policies, or in immigration legislation. Increasing hostility towards minorities like Roma and Sinti, or towards other groups designated as foreigners or outsiders, has made inroads on political discourse as well as political action: acts of terrorism and violence have been directed against such groups, and have become symbolically anchored in public consciousness by the name of Oberwart, Province of Burgenland (cf. Baumgartner and Perchinig 1995, 524). Boundaries between democratic parties and New Right movements are shifting; it has been easy to fuse xenophobia and Austrian identity symbolically, as in the slogan "Austria first!" employed by Freedomites in 1993, which aimed at rallying support for a "public initiative against foreigners". It seems, as ever, temptingly easy to be guided by stereotypes, personalization, simplification, and sharply drawn schemes of black and white, since this panders to the emotional needs of certain social groups.

Further enlightening examples of this aspect of Austrian political culture are such discussions as the one instigated by Freedomite leader Jörg Haider[5], who has declared the Austrian nation to be a *Mißgeburt* or "born deformed" (Scharsach 1992, 53). In its stead, Haider has been propagating a "Third Republic" which by and large aims at dismantling representative democracy. It should also be mentioned in this context that Austrians have proved somewhat lacking in self-confident citizenship (Hanisch 1991, 19).

In keeping with the general trend towards the use of symbolic forms in politics, Austria's Freedomites may serve to illustrate how (neo)populist parties and politicians use a simple binary symbolism following the basic pattern of "we" and "the others". Such communication through symbolic forms reduces complex political problems to "simple solutions" and feeds back into political culture, which – as a product of collective experience – is lent expression in political myths, rituals and symbols, and disseminated via mass media and their representation of reality.

In a comparative perspective it might, and should, be asked whether similar (or different) social or political taboos (or broken taboos) exist in other countries, how they can be described, and how their impact on political culture and political discourse may be assessed. Beyond that it is necessary to look for changes in political culture that may or may not have occured during recent years. After all,

> political culture is transmitted from generation to generation, but it is not transmitted unchanged, nor is it transmitted without question or by chance (...). Cultural transmission (...) is a lively and responsive thing that is continually being negotiated by individuals (Thompson, Ellis, Wildavsky 1990, 287).

Notes

1 "Die Hervorbringung und Aufrechterhaltung politischer Alltagskultur ist zunächst als soziale Konstruktion politischer Selbstverständlichkeit zu verstehen, welche die Mitglieder der sie erzeugenden und benutzenden Ethnie teilen ..." (Patzelt 1987, 278; translation by the author).

2 "ethnomethodologisch wären die szenischen Praktiken der Benutzung, Konstruktion oder Dekonstruktion politischer Selbstverständlichkeit (...) herauszuarbeiten" (Patzelt 1987, 278; translation by the author).

3 "Waldheims wenig überzeugende Rechtfertigung, er hätte nur seine Pflicht getan, ließ die Auffassung an Glaubwürdigkeit verlieren, daß Österreich das erste Opfer Hitlers gewesen sei, und stempelte diese Haltung zu einer opportunen Teilwahrheit der Generation des Staatsvertrags. Waldheims Diktum von der Pflichterfüllung für die deutsche Wehrmacht lenkte (in Form einer unbeabsichtigten Nebenfolge) die Aufmerksamkeit der Öffentlichkeit auf die Frage der Mitwirkung von Österreichern an den Verbrechen im Zweiten Weltkrieg. Zwar war die als Schlüsselaussage bewertete Äußerung Waldheims authentischer (für die Einstellung der Kriegsgeneration) als andere Erklärungen, sie stand aber in krassem Widerspruch zu Staatsräson und Gründungsdoktrin der Zweiten Republik" (Gehler and Sickinger 1995, 680; translation by the author).

4 Incidentally, the "Annexation Monument" (Anschlußdenkmal) in Oberschützen in the province of Burgenland has yet to be removed.

5 The Austrian Freedom Party (FPÖ) recently changed its name to "Freedomite movement".

References

Almond, G. A. and Verba, S. (1963), *The Civic Culture. Political Attitudes and Democracy in Five Nations*, New York.

Almond, G. A. and Verba, S. (1980), *The Civiv Culture Revisited*, Boston.

Bailer, B. (1993), *Wiedergutmachung kein Thema. Österreich und die Opfer des Nationalsozialismus*, Vienna.

Baumgartner, G. and Perchinig, B. (1995),'Vom Staatsvertrag zum Bombenterror. Minderheitenpolitik in Österreich seit 1945.' in Sieder, R. et al. (ed.) (1995), pp. 511–524.

Bell, D. (1973), *The Coming of Post-Industrial Society*, New York.

Bettelheim, P. and Harauer, R. (ed.) (1994), *Ostcharme mit Westkomfort. Zur politischen Kultur Österreichs*, Wien.

Dachs, H. et al. (ed.) (1991), *Handbuch des politischen Systems Österreichs*, Vienna.

Dachs, H. (1995), 'Von der Sanierungspartnerschaft zur konfliktgeladenen Unübersichtlichkeit. Über politische Entwicklungen und Verschiebungen während der Großen Koalition 1986–1994', in Sieder, R. et al. 1995, pp. 290–303.

Douglas, M. and Wildavsky, A. (1983), *Risk and Culture*, Berkely, University of California Press.

Edelman, M. (1964), *The Symbolic Uses of Politics*, University of Illinois Press.

Fessel+GFK (1987, 1992), *Austria Lifestyle Study*, Vienna.

Freiheitliche Partei Österreichs (ed.), *"Österreich zuerst"*. *Neue Argumente zur Ausländerpolitik*, Vienna.

Garfinkel, H. (1984), *Studies in Ethnomethodology*, Polity Press, Cambridge.

Gärtner. R. and Rosenberger S. (1991), *Kriegerdenkmäler*, Innsbruck.

Geertz, C. (1973), *The Interpretation of Cultures. Selected Essays*, Basic Books, New York.

Gehler, M. and Sickinger, H. (1995), 'Politische Skandale in der Zweiten Republik', in Sieder, R. et al (ed.) (1995), pp. 671–683.

Gerlich. P. (1989) 'Die Kehrseite der politischen Kultur – Tabus in der österreichischen Gesellschaft', in Heinrich et al. (1989), pp. 10–19.

Gerlich, P. (1991), 'Politische Kultur der Subsysteme', in Dachs, H. et al. (1991) pp. 457–465.

Gerlich, P. and Neisser, H. (ed.) (1994), *Europa als Herausforderung. Wandlungsimpulse für das politische System Österreichs,* Vienna.

Gesellschaft für politische Aufklärung (ed.) (1994), *Zur politischen Kultur Österreichs. (Dis)Kontinuitäten und Ambivalenzen,* Innsbruck.

Gibbins, J. R. (ed.) (1989, 1990), *Contemporary Political Culture. Politics in a Postmodern Age,* London, etc.

Halbwachs, M. (1950), La mémoire collective, Paris.

Hanisch, E. (1984) 'Politische Überhänge in der österreichischen politischen Kultur', *Österreichische Zeitschrift für Politik,* no. 1.

Hanisch, E. (1991), 'Kontinuitäten und Brüche: Die innere Geschichte', in Dachs, H. et al. (ed.) (1991), pp. 11–19.

Hanisch, E. (1994), *Der lange Schatten des Staates. 1890–1990,* Vienna.

Hanisch, E. (1995), 'Politische Symbole und Gedächtnisorte', in Tálos, E. et al. (ed.) (1995), pp. 421–430.

Heinrich, H.-G., Klose, A. and Ploier, E. (ed.) (1989), *Politische Kultur in Österreich,* Linz.

Heinrich, H.-G. (1990), 'Die österreichische Gesellschaft und die Mitteleuropa-Idee', in Pribersky, A. (ed.) (1990), *Europa und Mitteleuropa. Eine Umschreibung Österreichs?,* Vienna, pp. 46–58.

Heinrich, H.-G. and Wiatr, S. (1991), *Political Culture in Vienna and Warsaw,* Boulder.

Hunt, L. (1984) *Politics, Culture, and Class in the French Revolution,* University of California Press, Berkeley and Los Angeles

Inglehart, R. (1977), *The Silent Revolution,* Princeton.

Kaindl-Widhalm, B. (1990), *Demokraten wider Willen? Autoritäre Tendenzen und Antisemitismus in der 2. Republik,* Vienna.

Kerschbaumer, G. and Müller, K. (1992), *Begnadet für das Schöne: Der rot-weiß-rote Kulturkampf gegen die Moderne,* Vienna.

Knight, R. (1988), 'Ich bin dafür, die Sache in die Länge zu ziehen. Die Wortprotokolle der österreichischen Bundesregierung von 1945 bis 1952 über die Entschädigung der Juden', Frankfurt.

Kos, W. (1994), *Eigenheim Österreich. Zu Politik, Kultur und Alltag nach 1945,* Vienna.

Manoschek, W. (1995), 'Verschmähte Erbschaft. Österreichs Umgang mit dem Nationalsozialismus 1945–1955', in Sieder, R. et al (ed.) (1995), pp. 94–106.

Mattl, S. (1995), 'Vor der IV. Republik. Politische Kultur in Österreich im 20. Jahrhundert', *Zeitgeschichte,* no. 22, pp. 30–45.

Meissl, S., Mulley, K. and Rathkolb, O. (ed.) (1986), *Verdrängte Schuld – verfehlte Sühne. Entnazifizierung in Österreich 1945–1955,* Vienna.

Mörth, I. and Fröhlich, G. (ed.) (1994), *Zur Kultursoziologie der Moderne nach Pierre Bourdieu,* Frankfurt.

Patzelt, W. (1987), *Grundlagen der Ethnomethodologie,* Munich.

Pelinka, A. (1985), *Windstille. Klagen über Österreich,* Vienna.

Pelinka, A. and Weinzierl, E (ed.) (1988), *Das große Tabu – Österreichs Umgang mit seiner Vergangenheit,* Vienna.

Pelinka, A. (1994), 'Europäische Integration und politische Kultur', in Pelinka, A., Schaller, C. and Luif, P. (1994), *Ausweg EG? Innenpolitische Motive einer außenpolitischen Umorientierung*, Vienna, pp. 11–26.

Pelinka, A. (1995), 'Konkordanzdemokratie am Ende? Zur neuen Unübersichtlichkeit der politischen Kultur Österreichs', in Bundesministerium für Unterricht und Kunst (ed.) (1995), *1945–1995. Entwicklungslinien der Zweiten Republik*, Vienna.

Plasser. F. and Ulram, P. (1991), *Staatsbürger oder Untertanen? Politische Kultur Deutschlands, Österreichs und der Schweiz im Vergleich*, Frankfurt.

Plasser. F. (1994), 'Politische Modernisierungskonflikte in den 90er Jahren', in Gesellschaft für politische Aufklärung (ed) (1994), pp. 69–83.

Rathkolb, O. (1991), *Führertreu und gottbegnadet. Künstlereliten im Dritten Reich*, Vienna.

Reinprecht, C. (1992), *Zurückgekehrt. Identität und Bruch in der Biographie österreichsicher Juden*, Vienna.

Reiterer, A. F. (1988), *Nation und Nationalbewußtsein in Österreich. Ergebnisse einer empirischen Untersuchung*, Vienna.

Ringturmgespräche (1995), 'Inventur 45/55', 29 May to 1 June 1995, Vienna.

Scharsach, H.-H. (1992), *Haiders Kampf*, Vienna.

Schediwy, R. (1992), 'Die Waldheimat und ihre Kritiker – ein Rückblick', *Was – Zeitschrift für Kultur und Politik*, no. 68, pp. 53–62.

Seiter, J. (1995), 'Vergessen – und trotz alledem – erinnern. Vom Umgang mit Monumenten und Denkmälern in der Zweiten Republik', in Sieder, R. et al (ed.) (1995), pp. 684–705.

Sieder, R., Steinert, H. and Tálos, E.(ed.) (1995), *Österreich 1945–1995, Gesellschaft-Politik-Kultur*, Vienna.

Sternfeld, A. (1990), *Betrifft Österreich*, Vienna.

Stiefel, D. (1981), *Entnazifizierung in Österreich*, Vienna, Munich, Zurich.

Tálos, E., Dachs, H., Hanisch, E. and Staudinger, E. (ed.) (1995), *Handbuch des politischen Systems Österreichs. Erste Republik 1918–1933*, Vienna.

Thompson, M., Ellis, R. and Wildavsky, A (1990), *Cultural Theory*, Boulder, San Francisco, Oxford.

Uhl, H. (1994), 'Erinnern und Vergessen', in Riesenfellner, S. and Uhl, H. (ed.) (1994), *Todeszeichen. Zeitgeschichtliche Denkmalkultur in Graz und in der Steiermark vom Ende des 19. Jahrhunderts bis zur Gegenwart*, Vienna.

Ulram, P. (1990), *Hegemonie und Erosion. Politische Kultur und Wandel in Österreich*, Vienna.

Ulram, P. (1991), 'Politische Kultur der Bevölkerung', in Dachs, H. et al. (ed) (1991), pp. 466–474.

Vivelo, F.-R. (1988), *Handbuch der Kulturanthropologie*, Munich.

Voigt, R. (ed.) (1989), *Politik der Symbole – Symbole der Politik*, Opladen.

Wodak. R. et al. (1990), *"Wir sind alle unschuldige Täter". Diskurshistorische Studien zum Nachkriegsantisemitismus*, Frankfurt.

Ziegler, M. and Kannonier-Finster, W. (1993), *Österreichisches Gedächtnis. Über Erinnern und Vergessen der NS-Vergangenheit*, Vienna, Cologne, Weimar.

16   Russian and Central European Political Culture

Hans-Georg Heinrich

In search of the obvious?

The title of this essay has a somewhat dangerous, if popular, connotation. From a Central European perspective, it may seem as though Russian political culture were something remote and distinct from the culture of Central Europe, which claims the ancestry of Greek, Roman and Western European humanist, democratic, and urban civilization. Latin-Roman-Catholic civility and Eurasian orthodoxy thus appear juxtaposed, and the rift between Eastern and Western Roman Empires that occured in the 10th century is, as it were, reiterated at the end of the 20th. It is a fact that Eastern European elites have always looked westward for models of behavior and style, and that they have despised their Eastern neighbours as backward savages. This is particularly true of Eastern attitudes towards Russians. Not unlike panslavism, the frequently invoked "orthodox axis" represents a pragmatic myth rather than a reliable partnership. To speak of "panslavic" ties altogether shifts the argument into the realm of the imaginary. Poles regard Russians as their traditional enemies, and Czechs, Slovaks, and Croats have, to say the least, mixed feelings about Russians. When it comes to practical politics, all these myths have little explanatory value. In contrast to orthodox and panslavist rhetoric in the press and the State Duma (Russia's parliament), Russian industry has delivered military hardware to both Serbs and Croats in the Yugoslav war, and the fact that Russian and Ukrainian volunteers fight on the Serb side can mainly explained by the presence of illegal Serb recruitment offices in St. Petersburg and Kiev.

Throughout history, Central Europe has certainly been more open to Western influences than Russia. Russian policies toward its Western neighbours show a pattern of jagged oscillation between cautious opening towards, and almost total isolation from, the "sinful" West. As will be discussed below, this pattern reproduces itself in modern Russian politics. Comparative opinion polls covering the whole region show a clear West-East gradation with respect to many key items. Yet, leaving aside political correctness, it would be erroneous to draw the all too obvious conclusion that civility is oriented on the West-East vector, and that political culture becomes more European the farther West we go. First, many Russian respondents claim their allegiance to "Western" values; secondly, and perhaps more importantly, this view leads to the perilous conclusion that only an inculcation of Western values could turn Russia into a "normal" country. We need to accept the fact that in spite of the universality of values such as peace, happiness and self-respect, not all peoples pursue the same goals with the same

223

intensity. The words may be the same, but they mean something different in a different cultural context.

The same may be said of the term "political culture" itself. The battles raging since the 1950s on its "correct" use have obviously ended in a ceasefire and tacit agreement that it is to refer to long term attitudes towards politics. It is accepted as the residual factor space left by socio-structural and situational constraints. In other words, we should identify those attitudes and strategies which are preferred as problem-solving responses in a wide range of different contexts.

Comparing and contrasting different cultures also involves identifying certain patterns as "dominant" or characteristic of a given society. Cultural theory has studied the relationship between what were formerly called dominant cultures and subcultures, and has highlighted their mutual contingency. It describes cultures as a complex network of different "ways of life" representing different basic strategies of coping with life (Thompson et al., 1990). Comparative cultural research suggests, however, that certain ways of life are more characteristic of a particular society than others; there seem to be specific strategies preferred by most actors under different situational constraints.

When looking for the "typical Russian" or the "typical Central European" we need to be aware that we are talking about preferential strategies (not necessarily preferred outcomes) of a group which hardly lends itself to quantification at all: even if only 30% of a given population displays a certain pattern of behavior, the latter may nevertheless be perceived as dominant, since it may have acquired model character or differ sharply from standards accepted in the oberserver's community. Other strategical preferences may exist, but perhaps do not match preconceived expectations. We can assume something like a societal default strategy conforming to standards of normalcy in a given society. Selecting another available strategy is more costly and therefore rarer to be found. Default strategies make sense only in the context of a specific society, and are therefore also referred to as the logic of a system. Jokes touching on typical features of a national culture capture this element with precision: The Russian muzhik who is asked by the fairy to state his last wish and answers that he would like to see his neighbour's cow drop dead behaves according to received stereotype but also according to the intrinsic logic of Russian peasant society, where poverty is a means of beating a system based on exploitation and dependence. In a poor society of equals, exploitation and totalitarian rule are neutralized, while successful social climbing sets a dangerous precedent and invalidates that strategy. Hungarian agricultural workers, as Gyula Illyés observed in his classic piece of belletrist sociology, "The People of the Puszta" were perceived as lazy by their employers, but their go-slow tactics actually prevented total physical exhaustion and thus followed the logic of survival.

The Russian cultural syndrome

The genesis of a distinctly "Russian" cultural pattern can be related to typical contexts likely to dominate society throughout the lifespan of a generation. Russian society was militarized in the 13th century and has preserved this feature to this day. As some observers have pointed out, it was not the number of conflicts that made Russia different from other European nations. Originally, the Russians inhabited the dense forests of Northern Russia and came into conflict with their Southern neighbours when they began to migrate in search of arable land (Pipes 1978). This military confrontation brought them to the verge of extinction. It was a confrontation between military superpowers relying on different military strategies and correspondingly incommensurate social and political structures. The Russians, hard-pressed by permanent incursions of Mongols and Tatars, adapted their social system to that of the enemy as a survival strategy (Gumilev 1992). Society was militarized and centralized, the Tsar transformed into a Tatar Khan or Indian Mogul.

Traditional Russian political culture was shaped by the necessity to survive in the face of nature's forces, formidable and merciless enemies, and oppressive authorities. The average Russian was exposed to the whims of nature and politics; but he also received protection from the strong leader at the top. The resulting setup differed from Western European feudalism in that the the nobility's autonomy was severely curtailed, that the cities never achieved the status of a third, balancing power, and that the whims of the Tsar were accorded priority over the law. Arbitrary rule, extreme dependence on authorities and feudal lords, widespread distrust and lack of legal protection became ubiquitous features of the political system.

Especially since the end of Communist rule, many authors have argued that Russia has had ample experience with democracy (e.g. Rybkin 1994). Some cities, such as Novgorod Veliki, adopted the city statute of Magdeburg and enjoyed self-rule for more than 200 years; and Piotr Stolypin ended the medieval *mir* setup which had forced all villagers into the role of collective taxpayers or pre-Soviet *kolkhozniks*. Yet Moscow did not end self-rule in the Western cities and impose itself as imperial and orthodox center for nothing. Several hundred years later, Stolypin's brave attempt at creating an enclave of free enterpreneurship (roughly one third of Russia's regions implemented the reforms) grinded to a halt amid social unrest and political turmoil. It is no coincidence that Russia's potential for democracy and a market economy has never materialized. The Soviet system, as much as it was a caricature of Russian feudalism, was not grafted onto Russia by alien agents, Western ideologists, or other sinister forces. Even the fight against religion has antecedents in pre-Soviet history.

The Soviet system did not create Soviet man from scratch. It reinforced traditional features of Russian culture and refuted its own claims by its results. To be sure, it showed an impressive industrial and military buildup and channelled millions of peasants into the growing cities. The attempt to form a politically reliable new elite eliminated illiteracy. But the drive toward industrial modernization actually blocked social modernization, preserving attitudes typical of Russian peasant society: values and behaviour patterns such as lack of respect for the law, poor work ethics, fatalist egalitarianism, and endurance are paired with curiosity, friendliness, passion and a strong sense of social justice. In an appropriate context, they can still be found: in the depressing industrial deserts of large cities, in run-down kolkhozes, and in small isolated villages.

Today, Russia is still a militarized society, although the nation's political priorities were reversed 10 years ago. The only institutions which enjoy at least some credibility are the army and the Church. This is so despite the "invincible Red Army's" bad showing, its moral decay and its facing an economic impasse. The average Russian still grows up in quasi-military contexts. A childhood spent in conditions of closely circumscribed freedom is no exception, since the average family is structured around an authoritarian father (whose authoritarianism is usually aggravated by his wife's fear). Child beating is the rule even in overprotective and emotionally attentive, urban

225

intellectual families. Children are not usually asked what they want, but are expected to conform to their parents' wishes, to be eager helpers and diligent pupils, in short: obedient little soldiers. "Nachal´stvo znaet lutche" ("Your superiors know better") is the principal rule in a military context, but it applies equally to families, schools, and the workplace. It should not come as a surprise that precisely such strict rules and hierarchies tend to generate the exact opposite of what they are designed to achieve. Fierce individualism and radical anarchism have found fertile ground in Russia. When physical escape (such as migration to Siberia or the Southern marshes) was impossible, people chose inner emigration as a means of resisting outward pressure.

The lack of tolerance concerning groups with different political agendas has often been noted as a characteristic feature of Russian political culture. It is a feature inscribed in the military setup of society. An army in battle cannot afford long discussions on best solutions; it has to rely on the supreme wisdom of its leader. In Russian history, monolithic ideologies have always played a major role, be it in the form of religious orthodoxy or Marxism-Leninism. Democracy and its principle of competition between various views and strategies are disturbing idea for many Russians, who long for the comfort and security of the one "correct" world view. President Yeltsin represents a generation of Soviet leaders trying to substitute the "correct" ideology for the "wrong" one. He has certainly displayed more tolerance than his predecessors, but he is at the same time absolutely confident that his strategy is as true as his opponents' is false.

Exoneration from responsibility constitutes the reverse side of this military syndrome. Commanders and leaders are here to issue orders, and lowly soldiers must implement them no matter how irrational they might seem. "This is what nature wanted. Why? It is not our concern" runs a line in a song by Bulat Okudzhava, a Georgian bard who cast his poems into simple tunes virtually everyone knows, as the lyrics accurately reflect social and psychological realities "Ty nachal'nik ya durak – ya nachal'nik ty durak" (You are the commander, I am the fool, I am the commander, you are the fool): this popular quip aptly expresses the dynamics of subordination and exoneration. Russian emigrés in the West often have problems coping with paperwork and the multitude of official documents which define the bureaucratic existence of individuals in these societies. In Russia, paperwork and the handling of such documents is taken care of by faceless and anonymous bureaucrats. Until recently, the individual was not asked to make decisions or assume responsibilities for him- or herself.

Yet another aspect of the relationship between superiors and underlings is resistance, which is found primarily in covert forms such as go-slow strikes, sloppy work or cheating. Russian and Soviet military commanders did not enjoy a reputation of caring for their men; the Army was as repressive as the political system, factories and kolkhozes. Add dismally low wages to repression in the workplace, and you end up with what may be characterized by the "golden rule" of Soviet work: Workers pretend to work, and the state pretends to pay them. Many features and behaviour patterns in contemporary Russia can be explained by the unbroken vigour of this rule. When the Soviet Army supplied soldiers as cheap labour for the Hungarian *Ikarusz* bus factory's assembly lines during the final year of Soviet occupation, Hungarian foremen were astonished to find that the Russians fulfilled the daily requirement by noon, after which they hung around playing cards, smoking and drinking. Eager to apply the apparently revolutionary Soviet work organization, Hungarian inspectors found that the soldiers were using sledgehammers instead of screwdrivers to assemble bus components. The busses, incidentally, were to go to the Soviet Union and would have meant a grave hazard had they passed Hungarian clearance. Luckily, they did not, and it is questionable whether a Soviet or Russian inspector would have displayed equal vigilance. The Chernobyl accident is, of course, yet another illustration of the Russian/Soviet military syndrome. Poor work ethics are perilous in hazardous production environments, and it matters little that they are to be interpreted as acts of social resistance. But if leaders do not display concern for their people, how can the latter be expected to care about the former, or, for that matter, "the public good"?

There are other identifiable historical roots of sloppy attitudes to work. Constant military conflict, and an entire society in a frontier situation, did little to promote a culture of stability. The whole population was permanently on the move, fleeing military attacks, hunger and persecution by their own leaders. Ivan the Terrible is reported not to have known where his people were. Russian civilization was geared towards survival as the central value, and the artefacts it produced had a predominantly provisional character. For instance, Russians mainly used wood available in abundance to build their homes. In case of an enemy attack, this had the additional advantage that houses and belongings could be burned quickly so as not to afford an enemy shelter from long and cold winters. The burning of Moscow during Napoleon's campaign is an example of this strategy. But the many incursions also thwarted careful design and stable construction as much as it frustrated accumulation of durables in general. Stalin's great transformation, which relocated millions of people across the country, could still draw on the building traditions of a garrison state: housing was largely provisional, and instant solutions were given preference over sophistication. Having any kind of apartment at all was more of a concern than its style and finish, and the barracks atmosphere prevailing in the faceless and carelessly built socialist apartment blocks is ample proof of continuity in this respect. To an extent, such military virtues are presently being revived against the backdrop of a rapidly swelling internal migration, while construction of private homes or weekend cottages (dachas) is at the same time progressing at an unprecedented pace. The advent of stone construction has mostly been coeval with the coming of internal peace and a degree of political stability. Large-scale construction of family homes would thus both indicate and in turn promote political stability and prosperity, since private property rarely fails to appeal to owners' maintenance instincts.

Russians' sense of privacy and individualism has, however, developed to a much lesser extent than in Western societies. Acquisition of property was not encouraged by rulers, and as a legal institution, private ownership was undermined by the general absence of the concept of rights. The Russian nobility was not granted the right of self-government until the end of the 18th century and through an act of grace issued by Catherine the Great. In practice, this meant the right to organize local administration, but it did not make the nobility a countervailing power. There was no Magna Charta from which basic rights and freedoms could have been derived. Russian nobles did have power, but they had no rights (Simon 1995, 15).

This pattern has dominated for centuries. Russian society has displayed an astonishing capacity of self-organization from the beginning of the 20th century to the Stalinist period, and again during the past ten years. A major problem, however, has been the institutional weakness and poor credibility of legal guarantees extended by the state. Russia has witnessed periods of a strong and oppressive state, and others during which that state withdrew, decayed and left society largely to itself. But no matter how strong, the state has never been regarded as an institution upholding the law. On the contrary, it has the reputation of serving as a vehicle for the interests of powerful groups, and of deserving neither trust nor active support and participation.

Against this backdrop, the public/private dichotomy acquires its distinct features. In default of viable and credible legal guarantees of privacy, the sphere of private activities is deliberately neglected and left a rudiment reflecting overall insecurity and instability , and which has begun to resemble the public arena. Personal trust divides public and private spheres, and there is no generalized, abstract trust in society at large. Trust is a rare commodity, and is reserved for a select circle of family members and close friends. As public and private are increasingly merging, however, certain norms characteristic of public life have definitely entered the private sphere, such as reckless und rugged individualism, irresponsible self-seeking and the erosion of confidence in others.

Having introduced the unofficial privilege of enrichment for elites, the Brezhnev regime also had to tolerate private activities. This led to the privatization of public life, to the emergence of roped parties and large-scale corruption, and to the all but complete loss of what had remained in terms of respect for public and national interests. It was, by and large, the enriched elites and those waiting to enjoy their privileges on the basis of legal guarantees who initiated the drive towards democratization. While Russia may today well be, formally, a state ruled by law, the absence of political stability and of a legally constituted and secure civil society render formal constitutional guarantees worthless. This explains the tenacity of old habits in the new democratic context. It explains the virtual impossibility of reforming the political and economic system within a short term range. Market bolshevism of the Gaidar type was a traditional reform strategy of many Russian and Soviet leaders who faced inert masses and decided to push them a little by cracking the whip. Society, however, resorted to its time-honored arsenal of responses developed as survival strategies against repressive rulers. The reaction of industrial managers to the Gaidar reform is a case in point: the roots of their response touched layers much deeper than Communist ideology, which had, in any case, given way to pragmatism long ago.

Since the 1960s, the Soviet economic system has transformed into a market of "administrative rights" as main commodities: of access, that is, to the nerve centres of administration in the Ministries. As a vertical system in which incumbents of distributive positions set the basic parameters, it was eliminated in 1991. The only structures remaining therafter have been personal connections, which have so far held the economy together. The Boieva-report (a study of new "capitalist" management based on 220 in-depth interviews conducted in 1992) revealed that general managers preferred to uphold their personal contacts and networks. Administrative rights have since been bought officially, though not on an anonymous market, and not from abstract partners. Managers try to form a kind of cooperative, in which the value of a product is partly determined by an individual's social standing and reputation. The high mutual indebtedness of Russian enterprises is not only a consequence of inadequate money supply, but also of this social phenomenon. In 1992, only three to five percent of all business transactions used new market institutions such as commodity exchanges (Kharkhordine 1994, 31). New enterpreneurs representing the interests of anonymous investors are perceived as crooks and adventurers. This is by no means untypical for the early stages of a market economy, but it is miles apart from a modern economy which can only work on the basis of generalized trust. Russian management culture shows the same features as societal culture in general. Solidarity is valued more highly than individual business acumen, and personal connections are considered more reliable than the abstract mechanisms of the law and the market.

One very important aspect of Russian political culture has been the lack of a clear-cut ethnic identity. Russians themselves are a motley conglomerate of various ethnic units; moreover, they integrated large Asian (Tatars-Mongoles, Finno-Ugric tribes) and European populations during their empire's expansion. The "Russian" elites were of European or Tatar origin and predominantly spoke French. Stalin succeeded in suppressing all nationalities and in making citizens entirely dependent on the state. As a result, ethnic boundaries were blurred and national awareness became a minor component in the makeup of Russian identity. Russian nationalism proper dissolved in a vague official doctrine of "Soviet patriotism". This doctrine, however, was based on settlement patterns providing for decisive Russian presence in all major industrial regions of the Soviet Union.

The breakup of the Soviet Union changed political and economic boundaries, but not the traditional Russian national consciousness linked to the defunct Union's ethnopolitical space. In contrast to the situation in Yugoslavia, Yeltsin and his entourage used Russian separatism in order to bring down the Communist Party. Besides that, Russia is, of course, much too powerful a factor to be neglected. The former republics, including the Ukraine, depend much more on Russia than vice versa. In the eyes of most Russians, Russia should reunite with the Ukraine, Belorussia and Northern Kasakhstan, but in no event by forceful means. Hardline ministers and army representatives intending to "bring home Chechnia in a little victorious war" had their own reasons for doing so, but they are in no way representative of a political culture emphatically rejecting wars but prepared to fight for a good cause. The relatively peaceful dissolution of the USSR, contrasted with the Yugoslav tragedy, shows that stronger civic traditions alone cannot prevent the escalation of conflicts between bureaucratic ethno-nationalisms seeking redistribution in their favour. Without a stable regional order, such conflicts are bound to lead to nationalist frenzy, wars and destruction of civility.

Political culture in Central Europe

The term "Central European culture" has so far mainly been used in two senses. One takes ideals and aspirations found among the Eastern European intelligentsia at face value; the other disregards particularities for the purpose of a broader comparative perspective. Central Europe as a cultural and political space includes territories that were once part of the Austro-Hungarian monarchy or the German Empire. Their administrative and military elites were coopted and integrated into the system through a network of Catholic elite schools (or, in Germany, given the opportunity to pursue business or administrative careers on an even more selective social basis, cf. Stimmer 1993). The expansion of the Austrian and German administrative systems resulted in a transfer of the respective administrative cultures to East Central Europe. In contrast to Russia, ECE societies had the sort of well-trained and organized professional bureaucrats that have proved a major prerequisite for any modern democracy. But, as well as being inscribed in tradition, the values and aspirations of these elite were often limited functionally as well as inscribed in tradition. Like Austrian and German elites at the end of the 19th century, they feared parliamentary democracy as a system that would upset the stable and manageable order. In their understanding, the term "democrat" was tantamount to the modern bugbear "communist". Their loyalties were dynastic and class-bound – a far cry from a modern republican administration.

Nevertheless, Central European societies were better prepared for the evolution of democracy than Russia. The aristocracy had established itself as a countervailing power, and its privileges had been laid down in compacts that could have become the nuclei of any modern *Rechtsstaat*. Educational institutions and the army of the Habsburg monarchy greatly contributed to the formation of a national intelligentsia. During the last quarter of the 19th century, an investment boom laid the foundations for economic development. Despite its strained relations with the Imperial Court, Bohemia took the lead among non-German nations. Its middle classes were mainly active in the service sector and in small businesses. Its aristocracy was of German origin (the Czech nobility having been liquidated after its abortive attempts at establishing an independent and irridentist Hussite government) and was less of an obstacle to economic progress than, say, Hungarian high aristocrats who stubbornly clung to feudal privileges.

Except for Bohemia and Moravia, however, the non-German speaking territories of the Monarchy showed a marked delay in modernization. Unresolved conflicts of nationality complicated the task, leading, at times, to paradoxical consequences. The Hungarian parliament, the *Országgyülés*, for example, was much more efficient in its decisionmaking due to its ethnic uniformity, while the Austrian *Reichstag* was frequently paralyzed by nationalist polemics and squabbles. Nevertheless, urban centers throughout the monarchy shared a set of basic values, they had a common political identity and common ideas about the ideal polity.

It is to this – premodern but stable – phase in the evolution of Central Europe that the intelligentsia of the region has been reverting since the 1960s. The desire to hearken back to an idealized non-Soviet past has eclipsed the fact that many parameters which had constituted islands of pluralism and relative tolerance before 1918 were transformed with the breakup of the Dual Monarchy. Central Europe between the Wars was certainly no place for tolerance. Scarcity of jobs and arable land created an army of unemployed *lumpen* proletarians, while hegemonial powers used ethnic conflicts as vehicles for their ambitions and designs. Moreover, the backbones of urban elites had been Jewish or German; physical annihilation or expulsion of these groups left Central European ideas without the individuals who had once seen to their implementation. Stalinist revolutions created their own intelligentsia from among the offspring of "reliable" classes, i.e. workers and peasants, but it took them several generations to develop a kind of "Socialist liberalism" resembling similar developments in the West (Hegedüs/Forray 1989).

In East Central European societies, totalitarianism ended earlier than in the Soviet Union. Attempts at total *Gleichschaltung* failed to produce uniformity. Even within communist parties, different behaviour patterns evolved and surfaced at various meetings with the Soviets (Hegedüs 1985, Horn 1991). In Poland, the terror regime imposed by Stalin met with resistance from high Party functionaries, and was watered down successfully (Toranska 1985, 345). The old and new urban middle classes of Budapest, Prague and Warsaw obtained the unofficial right to privacy (later extended to workers in state and collective enterprises as a *de facto* right to exploit their employer, the state) as the by-product of a new philosophy of post-Stalinist leadership, which implied a comprehensive bargain between rulers and the ruled. Informal recognition of the Party's leading role was granted in exchange for the right to be left in peace – resulting in what appears today as an *uncivilized society*[1]. Present generations of East Central Europeans have grown up in a type of socialism that rewarded irresponsible behaviour in the public sphere ("Who does not steal from the state steals from his family"). Certainly the drive towards democracy was inspired by many factors, but here, too, the major internal motivation seems to have been the interest of an enlightened Party nomenclature to legalize its privileges (Konstantinov 1994). The impact of human rights movements is, by and large, exaggerated (Molnár 1995, 108).

Civil society was strongest in social settings patterned, for one reason or other, after Western models. This was the case in Slovenia (Jalusic 1994), which undoubtedly benefited from its marginal position within Yugoslavia and its vicinity to Austria. To a certain extent, civil society was able to hibernate in Czechoslovakia, too. It received a fresh impulse from the "Party of the excluded" (liberal Party members who became political and social outcasts after 1968) and could rely on vivid recollections of people who had once experienced democracy and a market economy. A continuity, however fragile, between democratic traditions and post-socialism was thus assured despite the orthodoxy of communist leaders.

In the private sphere, parts of the historical heritage were preserved due to the relatively short period of Stalinism. Traditional values such as quality craftsmanship, diligence, and achievement survived in enclaves of privacy. The private sphere did not, as in Russia, turn into a mirror image of the public realm, but became a refuge of the good life and of individual identity. This is why Western living standards are still rare in Russia, but commonplace in posh upper-end districts of Central European cities. This is also why enterpreneurship is not rejected as an alien economic practice (in both Central Europa and Russia, opinion surveys have produced varying and contradictory results due mainly to varying degrees of hope and frustration), while the state is still seen as indebted to individuals. An active economic role of the state becomes more popular as economic difficulties connected with the reforms increase. In this respect, surveys are a good indicator of the economic plight of the population (for Slovakia, cf. Mihalikova 1995, 53). Rose's observations on the concentration of the active and educated in the new private sector (Rose 1993, Rose 1995) possibly indicates a distribution of cultures (as "ways of life"). It also duly reflects the interests at the origin of the 1989 changeover to a new system, which, at least during its early stages, serves mainly the interests of the active and informed.

The devastation of public space has been the main result of Real Socialism. In addition to that, the private sphere preserved traditional attitudes and structures such as authoritarianism and paternalism. While the private realm was exposed to the impact of market modernity in the West, in East Central Europe conservative attitudes harboured in this sphere helped freeze modernization of the public sphere as well. The weakness of feminist movements in this region testifies to this assertion. The divide between responsible behaviour towards the family and friends on the one hand, and arbitrary self-seeking in the public sphere on the other, further erodes the authority of the law. Rule of law once existed in Central Europe as a set of relatively simple rules which everyone, including bureaucracies, must obey. It presupposed a general societal consensus on basic values and on a preordained social order, a setup ill adapted to rapid modernization and social change. The transformation to a conflict culture and a corresponding legal system providing procedural guarantees instead of substantive rules was never accomplished.

As far as the interplay between institutions is concerned, East Central Europe can today be called a democratic region. Free elections are held, constitutional courts review politically sensitive cases, and constitutions are heeded. The difference between working democracies and uncivilized polities resides, however, in the legal traditions of state bureaucracies. Modern states are run by bureaucracies bowing to the rule of law. Division of power and other democratic institutions are designed to keep in check the bureaucracy's theoretical omnipotence and ubiquity. While Western bureaucracies have operated in the context of a civil society, those of East Central Europe have not (Arató 1992). But Real Socialism did not succeed in annihilating pre-Socialist legal traditions everywhere, nor entirely. Re-establishing a democratically controlled state bureaucracy will be the easier, the stronger these traditions and the more stable the political and economic environment. Societies which have never experienced regular bureaucratic procedures are least likely to succeed in setting up administrative bodies enjoying the trust of those affected by their decisions. They are bound, at least for some time, to remain uncivilized societies.

Empirical surveys testify to a lack of confidence in bureaucracies, as well as to continuities in traditional attitudes towards the law.(Ádam and Heinrich 1987, 142). In Poland, equality is generally considered more important than legality (Kourilsky 1994, 5f.; Heinrich and Wiatr 1990). Polish views on the role and function of law come close to traditional Russian attitudes, according to which "justice" (pravda) is a personal attribute or quality with strongly religious connotations. In both Russia and Poland, violation of the law is seen as a moral contravention, whereas procedural and institutional considerations dominate a country like France. In Poland, there is a particular fear of anarchy and chaos: reestablishing and maintaining order is seen as the function of law. In this respect, Poles approach a Western perspective. For Russians, the law is little more than a moral rule which ought not to be broken (Kourilsky 1994). Similarly, there is no concept of "liberty" in Poland and Russia, "svoboda" and "wolnosc" both denoting unlimited freedom rather than legally structured liberty with its concomitant civic counterweights attached. Freedom in this sense becomes an explosion or escapism. In general, the subordination of law to power in Russia and East Central Europe has meant that power became a survival resource, a fact left largely unchanged by its monetarization.

East Central Europan society is, in many ways, a premodern one. Socialist modernization has not been able to create generalized trust. Nor has it succeeded in equalizing trust levels between public and private spheres. On the contrary, socialism has created generalized distrust. Institutions of the public sphere (with the exception of the army and the Church) enjoy temporary confidence at best. Empirical diagnosis of trust in institutions and in the political system generally (i.e. in democracy as a value) shows a consistent West-East gradation. Western Europeans trust their courts and parliaments more than Eastern Europeans, while citizens in (post-)socialist countries feel that they can only survive with the help of strong networks colonizing bureaucratic institutions. These structures (from the mafia to local clans and various roped parties) have successfully been privatizing public institutions and their resources since the 1960s.

Apart from illegal counterorganizations, powerful and arbitrary bureaucracies also create weak individuals. US-Americans and most Western Europeans display a greater sense of political efficacy; they believe that the conduct of their rulers depends in part on themselves, they believe, in other words, in their power to influence political decisions. If we place old Anglo-Saxon democracies on the left end of the spectrum, Central European countries reach average scores, while a majority of Russian citizens (66% of Muscovites by February 1993) agree that "for people like you and me there is no possibility to influence government decisions" (Demidov 1994, 234). Interestingly enough, Austria has approximately the same rating as the Chech and the Slovak Republics. Regional cultures in Poland are another example of the longevity of behaviour patterns: during local elections in 1990, turnout ranged from 34 to 38% in the former Russian territories, and from 57 to 68% in Galicia, where Austrian rule (which was by no means only benevolent) had led to a well-entrenched system of self-government by the time of World War I (Buszko 1994, 751).

Summary and conclusions

Central European and Russian political cultures appear similar when compared to those of traditional and established democracies. The renewed popularity of former communist parties in East Central Europe testifies to the fact that Real Socialism was not just an alien, Russian-imposed system, but one which met objective needs for social security. The peasant masses of Eastern Europe and pre-revolutionary Russia had been passive and dependent, and had never experienced democratic participation. Similarly, the elites and masses of today were formed during the final stages of Real Socialism. All political groups, therefore, share certain characteristic features of the old system as well as of the each country's particularities. Government and opposition in Socialist Poland used to engage in violent rhetorics while displaying remarkable moderation in action. In Hungary, action and rhetorics became moderately pragmatic after 1962. Right-wing extremism is socially isolated, as is left-wing orthodoxy.

A majority of the population has preserved attitudes characteristic of later stages of Real Socialism. "Civic society" means uninhibited pursuit of private activities for many. "Democratic" values seem to be of little use for them, unless democracy is identified with a system improving living standards. For the elites, values such as "rule of law" are more important since in their eyes, democracy's *raison d'être* is to guarantee their position in society and as consumers.

When compared to the political culture of modern Russia, East Central European cultures seem more "European" and thus more ready to embrace Western concepts of government. The evolution of democracy, however, hinges on changes in the deep structures of society, on the equilibration of power balances, on the genesis of such civic virtues as tolerance or forbearance, and the existence of a republican civil service respectful of the law and resisting the pressures of its clienteles. Despite all claims to the contrary, uncivilized economies and polities still prevail in East Central Europe. Survey research on key items locates developed democracies on one end of a continuum, Austria close to post-Socialist societies and Russia linked to Western Europe via Eastern Europe. This can be explained by the heritage of various communisms prevailing in the respective societies, but also by examining their more distant past. East Central Europe established the preconditions for a civic culture prior to the arrival of socialism. In Russia, economic development partly surpassed Central Europe at the turn of the century, while social modernization had barely begun (Selunskaia 1993). Yet, these preconditions alone will not suffice to breathe new life into what had hibernated under Stalinist repression. The political culture of a 19th century *Rechtsstaat* is not yet compatible with that of an open and dynamic society.

Note

1   An uncivilized society may be defined as one whose legal system and institutions are have been appropriated by private interests. In the eyes of the citizens, impartial decisions are impossible; justice must therefore be sought via the perversion of official procedures.

References

Adám, A. and Heinrich, H.-G. (1987), *Society, Politics and Constitutions. Western and East European Views*, Böhlau, et al.

Arató, A. (1992), 'Social Theory, Civil Society and the Transformation of Authoritarian Socialism', in Gáthy, V. and Jensen, J.(eds.), *Citizenship in Europe?*, Szombathely, pp. 116–146.

Demidov, A. (1993), 'Die postkommunistische Gesellschaft Rußlands', in Plasser, F. and Ulram, P. (eds.) (1993), pp. 221–238.

Füstös, L. and Szakolcsai, Á. (1994), 'Value Changes in Hungary 1978–1993. Continuity and Discontinuity in East-Central European Transition', EUI Working Papers in Political and Social Science, European University Institute, Florence.

Gumilev, L. N. (1992), *Ot Rusi k Rossii, Ocerki etniceskoj istorii*, Moskva.

Hegedüs, A. T. and Forray, K. R. (1989), *Az újjáépítés gyermekei – a konzolidáció gyermekei*, Magvetö, Budapest.

Hegedüs, A. (1985), *Élet egy eszme árnyékában*, Vienna.

Heinrich, H.-G. and Wiatr S. (1991), *Political Culture in Vienna and Warsaw*, Westview Press, Boulder.

Horn, G. (1991), *Cölöpök*, Zenit Könyvek, Budapest.

Jalusic, V. (1994), 'Troubles with Democracy: Women and Slovenian Independence', in Benderly, J. and Kraft, E. (eds.) (1994), *Independent Slovenia. Origins, Movements, Prospects*, New York.

Buszko, J. (1994), 'Das soziale und politische Erbe im unabhängigen Polen', *Österreichische Osthefte*, vol. 36, 4, pp. 741–752.

Kharkhordine, O. (1994), 'L'éthique corporatiste, l'éthique de samostojatel'nost'et l'ésprit du capitalisme: réflexions sur la création du marché en Russie post-soviétique', *Revue d'études comparatives Est-Ouest*, no. 2, pp. 27–56.

Konstantinov, Y. (1994), 'Survival Strategies in Bulgaria: Patterns of Rearrangement in the Process of Post–89 Changes', Paper read at the Fifth Karl Polányi Conference, Vienna, 10 to 13 Nov.

Kourilsky-Augeven, C. (1994), 'Les individus, le sens de la justice et la loi (France, Pologne, Russie)', *Revue d'études comparatives Est-Ouest*, no. 3, pp. 5–13.

Mihalikova, S. (1995), 'Demokratikus politikai kultura-ábránd vagy realitás', *Politikatudományi Szemle*, no. 2, pp. 43–66.

Molnár, M. (1993), 'A demokrácia hajnalodik Keleten', *Politikatudományi Szemle*, no. 2, pp. 74–111.

Pipes, R. (1993), 'Rossiia pri starom rezhime', *Nezavisimaia Gazeta*, Moskva.

Plasser, F. and Ulram (1993), 'Zum Stand der Demokratisierung in Ost-Mitteleuropa', in Plasser, F. and Ulram, P. (eds.), *Transformation oder Stagnation. Aktuelle politische Trends in Osteuropa*, Signum, Vienna, pp. 9–88.

Rose, R. (1993), 'Contradictions Between Micro- and Macro-Economic Goals in Post-Communist Societies', *Europe-Asia Studies* vol. 45, no. 3, pp. 419–444.

Rose, R. (1995), *New Russian Barometer–IV*, Centre for the Study of Public Policy, University of Strathclyde, Preliminary Report.

Rybkin, I. P. (1995), *Die Staatsduma*, Jugend & Volk und Dachs-Verlag, Vienna.

Selunskaia, N. B. (1993), 'Rossiia na puti ot patriarkhalnosti k civilizacii', *Vestnik Moskovskogo Universiteta*, vol. 12 (1993), no. 6, pp. 13–19.

Simon, G. (1995), 'Zukunft aus der Vergangenheit. Elemente der politischen Kultur in Rußland', *BIOST*, no. 10.

Stimmer, G. (1993), 'Anstalts- und Bundelite als Grundtypen der politischen Elitenbildung. Elitenrekrutierung in Österreich von der Monarchie bis zur Zweiten Republik', unpublished *Habilitation*, University of Vienna.

Thompson, M., Ellis, R. and Wildavsky, A. (1990), *Cultural Theory*, Westview Press, Boulder.

Toranska, T. (1985), *Oni*, Aneks Publishers, London.

Appendix Survey Data

Table 1 Sense of political efficacy

("People like me have no influence on government")

country	percent disagreeing
USA (1988)	58
Switzerland (1991)	46
Australia (1987)	42
Netherlands (1988)	38
Germany (1989)	32
Austria (1989)	24
Czech Repuplic (1993)	23
Slovakia (1993)	22
Hungary (1992)	10
Poland (1992)	8

Source: Plasser and Ulram 1993, 44

Table 2 Authoritarian tendencies

% agreeing	"The country needs a strong man"	"Would welcome dissolution of parliament and abolition of parties"
Austria	22	8
Czech Republic	22	19
Slovakia	19	20
Hungary	26	24
Poland	36	40

Source: Plasser and Ulram 1993, p. 46

Table 3 Trust in institutions

					in percent of respondents with high trust	
	Switzerland	Austria	Poland	Hungary	Czech Rep.	Slovakia
army	49	41	63	40	42	52
church	–	–	50	42	33	45
police	55	51	30	34	36	35
law courts	52	50	32	40	38	37
Parliament	51	31	12	20	19	24
media	–	21	35	34	40	39

Source: Plasser and Ulram 1993, 59

Table 4 Functions ascribed to the state (Slovakia, 1994)

	in percent
employment for everybody	89
price regulation	80
provision of housing	75

Source: Mihalikova 1995, 53

238

Table 5 Political values in Russia

	percent agreeing
"It is most important to maintain silence and order in our society"	80.6
"It is most important to have personal liberty and the right to do what one wants without state interference"	19.4
"Our state must develop in the Western European style"	21.5
"Our state must develop according to our traditions"	78.5

"We do not need either parliament or elections, but a strong person who can take decisions and implement them fast"	in percent
absolutely correct	21.1
rather correct	22.1
rather incorrect	20.1
incorrect	20.4

Maximal trust in institutions and personalities	in percent
President Yeltsin	5.6
parties	1.4
law courts	3.8
government	2.3
army	14.4
parliament	2.5
church	22.7
private firms	3.2
foreign organizations and experts counseling the government	3.3